Geochemistry
of Epigenesis

Monographs in Geoscience

General Editor: Rhodes W. Fairbridge

Department of Geology, Columbia University, New York City

B. B. Zvyagin
Electron-Diffraction Analysis of Clay Mineral Structures–1967

E. I. Parkhomenko
Electrical Properties of Rocks–1967

L. M. Lebedev
Metacolloids in Endogenic Deposits–1967

A. I. Perel'man
The Geochemistry of Epigenesis–1967

In preparation:

A. S. Povarennykh
Crystal-Chemical Classification of Mineral Species

S. J. Lefond
Handbook of World Salt Resources

Geochemistry
of Epigenesis

Aleksandr I. Perel'man
Department of Geochemistry
Institute of Ore Deposits, Petrography, and Mineralogy
Academy of Sciences of the USSR, Moscow

Translated from Russian by
N. N. Kohanowski
Department of Geology
University of North Dakota, Grand Forks

With a Foreword by
Rhodes W. Fairbridge

℗ PLENUM PRESS · NEW YORK · 1967

Aleksandr Il'ich Perel'man, senior scientist in the Department of Geochemistry at the Institute of Ore Deposits, Petrography, and Mineralogy of the Academy of Sciences of the USSR in Moscow, was born in Moscow in 1916. In 1938, he was graduated from Moscow State University, as a specialist in topology, and received his doctorate in mineralogy in 1954. Perel'man is the author of seven books, many of which have been translated into Chinese, Hungarian, Bulgarian, and English, and 135 scientific papers. His research concerns mainly the geochemistry of landscapes and the geochemistry of epigenetic processes. In 1966, he was awarded a gold medal by the Geographical Society of the USSR for his work on the geochemistry of landscapes.

ISBN 978-1-4684-7522-7   ISBN 978-1-4684-7520-3 (eBook)
DOI 10.1007/978-1-4684-7520-3

The original Russian edition was published by Nedra Press, Moscow, in 1965.

Александр Ильич Перельман
ГЕОХИМИЯ ЭПИГЕНЕТИЧЕСКИХ ПРОЦЕССОВ

GEOKHIMIYA EPIGENETICHESKIKH PROTSESSOV

GEOCHEMISTRY OF EPIGENESIS

Library of Congress Catalog Card Number 65-25241

Dedicated to Alexander A. Saukov
An outstanding Russian geochemist

Foreword

In its classical sense "epigenesis" refers to all geological processes originating at or near the surface of the earth. It thus embraces all those phenomena which we associate with the landscape; Perel'man has already written extensively on this subject. The landscape, in the physical sense, is controlled by the interaction of exogenic and endogenic agencies—on the one hand, the atmosphere, the wind, the rain, and other components of the weather, the forces of running water and the planetary controls of gravitational and tidal nature; and on the other hand the materials of the earth's crust, from sediments to metamorphic rocks and igneous materials from deep endogenic sources.

In practical terms the epigene region involves the products of weathering, the soils, the transported material, the colluvium of hillsides, and the alluvium of stream valleys. It involves those landforms that are products of the erosional sculpturing of the landscape, as well as those that result from accumulation, such as glacial moraines and desert sand dunes. The science of geomorphology is gradually beginning to evolve from a passive cataloging of scenery and its deduced causes (in the Davisian sense) into a vigorous study of dynamic processes. These are partly geophysical, in the sense of hydraulics and mechanical studies, and partly geochemical. It is in regard to the latter that we owe a considerable debt to Perel'man and others of the Soviet school of geochemists, who are bringing a great deal of new and carefully analyzed data to bear to help explain the multitude of problems of the epigene zone.

Generally speaking, this is the zone of rainwater, well oxygenated, high in CO_2, and of low pH. At the ground surface, in the soil, in the groundwater, and deeper in the artesian circulation, the former rainwater goes through many changes: the pH may swing over to the high alkaline range, and the Eh to the reducing field. In regions of ore deposits it plays a critical role in transporting ions downwards, forming "supergene deposits." Perel'man finds this term useful in general application, so we have a "supergene zone." Those minerals concentrated by the surface waters are, in contrast, the "epigene deposits." The former are thus intimately related to the latter.

Seen from the viewpoint of the volution of sediments which begin their diagenetic evolution with syndiagenesis, engendered by the reactions appropriate to their site of origin, the end-stage in the evolution of a sedimentary rock is what this writer has called "epidiagenesis." In both the earliest and latest stage the environmental chemistry is dominated by its reactions with the biosphere. Thus, Perel'man in Chapter 1 begins with a discussion of "living matter." This should make salutary reading for some of our geologists who do not see the role of biology in our science.

Rhodes W. Fairbridge

New York City
August, 1967

Preface

Identical chemical phenomena are often observed in entirely unrelated rocks. For example, the weathering of feldspars and their subsequent conversion to kaolin and gypsum takes place in all types of rocks. We can thus speak of the geochemical types of the epigenetic processes which are common to all the formations of the supergene zone. The conclusion is inescapable that such epigenetic processes should be considered as a specialized group of phenomena. The methods of geochemistry lend themselves to the study of such processes.

A review of such epigenetic processes constitutes the subject of this book, which has grown about the nucleus of the lectures delivered by the author at the Institute of Ore Deposits, Petrography, Mineralogy, and Geochemistry (IGEM) of the Academy of Sciences of the USSR.

The geochemical approach to the study of epigenetic processes is indispensable to the solution of both theoretical and practical problems, such as geochemical exploration and theories of exogenous ore formation.

Some of the concepts treated in this book have been developed by the author during his investigations in Central Asia and other parts of the Soviet Union. The existing literature has been freely drawn upon.

In preparing this book for publication, the author had in mind the needs of geologists, who, working in various remote regions, are going to use geochemical methods of investigation, particularly methods for finding ores. To be useful to such persons, the material had to be presented in a systematic way and so some material had to be included which does not represent anything new.

The first edition of this book was published in 1961. This second edition was revised and enlarged. Valuable suggestions were made by Dr. A. A. Beus, Director of the IGEM, and by S. G. Batulin, O. I. Zelenova, and N. A. Volukhovykh. S. G. Baroyanits did a great deal of the editing.   The author is grateful to them all.

A. I. Perel'man

Contents

Introduction

In the supergene zone of the earth's crust, chemical elements migrate at low temperatures and pressures. This zone may extend for several thousand meters below the surface. It is underlain by a zone of metamorphism, which, in turn, is underlain by a zone of magmatism. The name "supergene" was coined by Fersman (1934), who considered it to be a zone of various surface and near-surface phenomena such as the formation of soils, weathering, sedimentation, etc. Here, the term will be used exactly in that sense.* It should be mentioned that the processes of the supergene zone are in contrast to those of the hypogene zone, which occur at high temperatures and pressures.

The continuous influx of solar energy onto the surface of the earth promotes numerous supergene processes and increases their complexities, energies, and intensities. Although temperatures and pressures are low, the migration of chemical elements in this zone is pronounced. Substances may become highly dispersed or segregated to an extent unknown in hypogene environments.†

Igneous, metamorphic, and sedimentary rocks, formed within this zone, change through the exchange of chemical components with the adjacent media. In relation to the affected rocks, such exchanges are definitely secondary and are not predetermined by

---

*The term "supergenesis" (= supergene processes) has also been used in other ways. For instance, it was used as a synonym for weathering. However, we prefer to use it in a broader sense and in agreement with the ideas expressed by Fersman, Saukov, and others.

†The supergene zone is characterized by enormous quantities of certain rocks, ores, and gases which consist of not more than three elements, e. g., $NaCl$, $CaSO_4$, CH_4, etc.

the original processes of rock formation. Being truly epigenetic,* they are not the inevitable results of diagenesis, magmatism, etc.

The formation of secondary dispersion haloes, which are the main objects of geochemical investigations, is entirely due to epigenesis. Various exogenes have been formed by epigenetic segregations of their ore elements. Examples are copper impregnations in sandstone, or deposits of sulfur or uranium.

So far, the study of epigenesis has not touched upon all its phases to the same degree. For instance, little is known about the chemical activity of water, while soil formation has received much attention.

In this book, epigenetic processes will be considered from the geochemical point of view. Our discussions will be based on the abundance concept, on laws of migration of the elements within the earth's crust, and on the characteristics of different geochemical landscapes.

The most important factor controlling the epigenetic alteration of rocks is chemically active water. Unlike the solutions we prepare in the laboratory, ground waters are biologic reagents in a sense that all the epigenetic rock alterations are conditioned by the presence of living organisms. It follows that all epigenetic rock alterations are conditioned by the complex of processes: biologic, mechanical, physical, and chemical.

As a rule, several different processes run concurrently, each being influenced and modified by the others. This is particularly true of the biologic processes. Nevertheless, each of those concurrent processes should be evaluated singly in order to correctly understand it and to classify it.

Our classification of these processes is based on the mode of migration of the matter. The principal types (Perel'man, 1961) are:

1. Biogenic migration
2. Physical and chemical migration
   a. Aqueous
   b. Aeolean

*The concept of epigenetic processes (= epigenesis) was developed as a part of the theory of mineralization. It is used by many geologists and geochemists. Unfortunately, the term has been given different meanings. Some petrologists designate as epigenetic those processes which immediately precede metamorphism. Drawing attention to this circumstance, we hope that the usage of this word will be soon defined in narrow terms.

3. Mechanical migration
   a. Aqueous
   b. Aeolean
   c. Gravitational

The chemical reactions of the supergene zone vary drastically from those conducted in the laboratory. The most important are reactions which cannot be duplicated in the laboratory. Among such nonreproducible factors are (1) the activity of living organisms and (2) the nonuniform distribution of chemical elements in the earth. To examine the effect of these factors on the migration of atoms is one of the objects of this book.

Chapter 1

Geochemical Role of Living Matter (Biogenic Migration)

It was Vernadskii who showed that living organisms are important agents controlling the migration of chemical elements within the supergene zone. It is the totality of all living organisms that makes an impression on the environment, and not any single living being. Vernadskii called this total the "living matter." One can speak of the living matter of all land or may limit the discussion to some smaller area — a hilltop or a stratigraphic suite. The total mass of living matter on the earth is many times smaller than the mass of rocks. However, living matter is being constantly created and destroyed, with the result that enormous quantities of atoms become involved in it.

The biological cycle exists within the earth's crust and consists mainly of the formation and destruction of organic substances (Fig. 1). This cycle regulates the geochemical features of the processes which go on within the supergene zone.

FORMATION OF LIVING MATTER

The formation of living matter from organic compounds existing in the environment is largely the result of the metabolism of green plants. These synthesize organic compounds from CO_2 of the air, water, and mineral salts. Photosynthesis may be explained by the following typical reaction:

$$6CO_2 + 6H_2O + 674 \text{ cal} \xrightarrow[\text{chlorophyll}]{\text{sunlight}} C_6H_{12}O_6 + 6O_2.$$

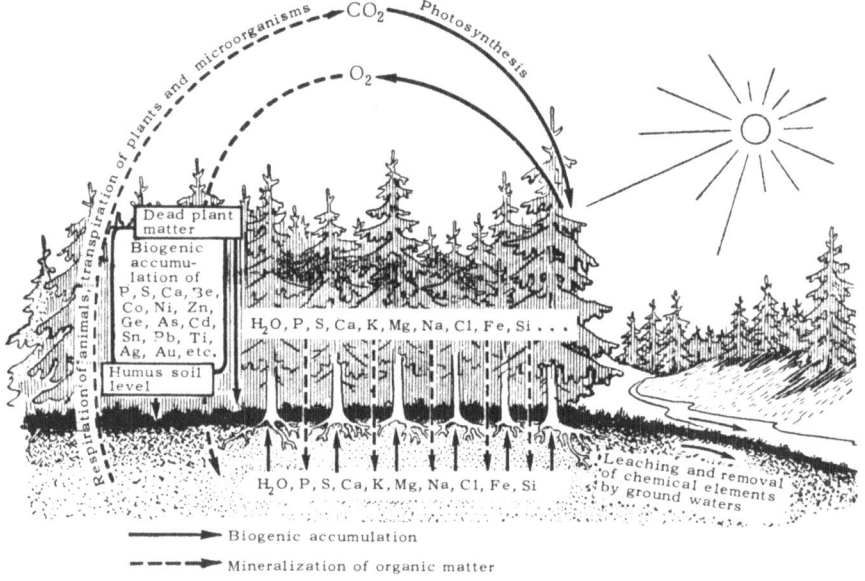

Fig. 1. Biologic cycle of elements on a landscape.

Thus, this carbohydrate is formed from six molecules of CO_2, six molecules of water, and 674 calories. Carbon dioxide is obtained from either the air or the soil, water from either the soil or some aquifer, and the calories from sunlight. The three materials are combined by the action of chlorophyll — the green pigment in plant leaves. Even such compounds as proteins are formed in this way. Calcium, magnesium, potassium, iron, and other elements are sorbed from soils to manufacture complex organometallics. Much organic matter, rich in energy, is thus accumulated at the earth's surface and in upper levels of the sea. Atoms of C, N, H, etc., which constitute living matter, may be regarded as "charged with energy."

The process described above is referred to as the "biogenic accumulation" of such compounds as H_2O, CO_2, or mineral salts.

The occurrence of elements in the form of organic complexes is a special mode of their occurrence within the earth's crust.

Animals and certain plants — for instance, fungi — as well as most microorganisms are not capable of organic synthesis from inorganic compounds. They build their tissues and organs by utilizing available proteins, fats, carbohydrates, etc. Thus, the original photosynthesis leads to a great variety of living matter.

Some microorganisms synthesize their organic compounds from mineral substances, not with sunlight but with energy liberated during some chemical reaction — e.g., the oxidation of iron. This chemosynthesis is also capable of producing living matter, although it is not common.

Although all chemical elements pass through the stage of incorporation within living matter, this is more typical of some elements than of others. In fact, much nitrogen, potassium, phosphorus, and even sulfur is constantly tied up in organic complexes.

Essentially, organisms consist of those chemical elements which occur either in the gaseous state (O_2, N_2, CO_2) or form water-soluble salts [NaCl, KCl, $Ca(HCO_3)_2$, etc.]. Oxygen is the most abundant in all organisms and is followed in descending order by C, H, Ca, K, N, Si, P, Mg, S, Na, Cl, and Fe.

Depending upon their abundance in living organisms, all chemical elements may be classified into macroelements, microelements, and ultramicroelements.

Knowledge of the composition of the ash of organisms is indispensable for a successful solution of geochemical problems,

Average Composition of Living Matter (after Vinogradov, 1954)

Macroelements ($n \cdot 10 - 10^{-2}$)

O — 70	N — 0.3	S — 0.05
C — 18	Si — 0.2	Na — 0.02
H — 10.5	Mg — 0.04	Cl — 0.02
Ca — 0.5	P — 0.07	Fe — 0.01
K — 0.3		

Microelements ($n \cdot 10^{-3} - 10^{-5}$)

Al — $5 \cdot 10^{-3}$	Zn — $5 \cdot 10^{-4}$	Pb — $5 \cdot 10^{-5}$
Ba — $3 \cdot 10^{-3}$	Rb — $5 \cdot 10^{-4}$	Sn — $5 \cdot 10^{-5}$
Sr — $2 \cdot 10^{-3}$	Cu — $2 \cdot 10^{-4}$	As — $3 \cdot 10^{-5}$
Mn — $1 \cdot 10^{-3}$	V — $n \cdot 10^{-4}$	Co — $2 \cdot 10^{-5}$
B — $1 \cdot 10^{-3}$	Cr — $n \cdot 10^{-4}$	Li — $1 \cdot 10^{-5}$
Rare earths — $n \cdot 10^{-3}$	Br — $1.5 \cdot 10^{-4}$	Mo — $1 \cdot 10^{-5}$
Ti — $8 \cdot 10^{-4}$	Ge — $n \cdot 10^{-4}$	Y — $1 \cdot 10^{-5}$
F — $5 \cdot 10^{-4}$	Ni — $5 \cdot 10^{-5}$	Cs — about $1 \cdot 10^{-5}$

Ultramicroelements ($< 10^{-5}$)

Se — $< 10^{-6}$	Hg — $n \cdot 10^{-7}$
U — $< 10^{-6}$	Ra — $n \cdot 10^{-12}$

Table 1. Average Contents of Ashes of Land Organisms with Their Coefficients of Biologic Absorption*

Chemical element	Abundance in lithosphere	Average content in ashes of land plants	Coefficient of biologic absorption
Ca	3.60	$n \cdot 10\ (n < 2)$	n
K	2.50	$n \cdot 10\ (n < 3)$	n
Si	29.5	n	$0.n$
Mg	1.87	$n\ (n < 5)$	n
P	0.093	$n\ (n < 5)$	$n \cdot 10 - n \cdot 100$
S	0.047	$n\ (n < 5)$	$n \cdot 10 - n \cdot 100$
Na	2.50	n	n
Cl	0.017	$n\ (n < 5)$	$n \cdot 10 - n \cdot 100$
Fe	4.65	$0.n - n\ (n < 2)$	$0.n - 0.0n$
Al	8.05	$0.0n$	$0.0n$
Ba	0.065	$0.0n - 0.n$	$0.n - n$
Sr	0.034	$0.0n - 0.n$	n
Mn	0.10	$\sim 1 \cdot 10^{-2}$	$0.n$
B	$1.2 \cdot 10^{-3}$	$0.0n - 0.00n$	$n - n \cdot 10$
La	$2.9 \cdot 10^{-3}$	$n \cdot 10^{-2} - n \cdot 10^{-3}$	n
Ti	0.45	$0.0n$	$0.0n$
F	0.066	$0.0n$	$0.n - n$
Zn	0.0083	$\sim 1 \cdot 10^{-2}$	n
Rb	0.015	$0.01 - 0.00n$	$0.n - 0.0n$
Cu	0.0047	$\sim 5 \cdot 10^{-3}$	$0.n$
V	0.009	$\sim 5 \cdot 10^{-4}$	$0.0n$
Cr	0.0083	$\sim 5 \cdot 10^{-4}$	$0.0n$
Cd	$1.3 \cdot 10^{-5}$	$n \cdot 10^{-4}$	n
Ge	$1.4 \cdot 10^{-4}$	$n \cdot 10^{-3}$	$0.n - n$
Ni	0.0058	$\sim 1 \cdot 10^{-3}$	$0.n$
Pb	$1.6 \cdot 10^{-3}$	$\sim 1 \cdot 10^{-4}$	$0.n$
Sn	$2.5 \cdot 10^{-4}$	$n \cdot 10^{-4}$	$0.0n - 0.n$
As	$1.7 \cdot 10^{-4}$	$5 \cdot 10^{-4}$	$n - 0.n$
Co	0.0018	$\sim 4 \cdot 10^{-4}$	$0.n$
Li	$3.2 \cdot 10^{-3}$	$n \cdot 10^{-4} - n \cdot 10^{-3}$	$0.n$
Mo	$1.1 \cdot 10^{-4}$	$n \cdot 10^{-4} - n \cdot 10^{-3}$	n
Y	$2.9 \cdot 10^{-3}$	$n \cdot 10^{-4} - n \cdot 10^{-3}$	$0.n$
Cs	$3.7 \cdot 10^{-4}$	$n \cdot 10^{-4}$	$0.n$
Se	$5 \cdot 10^{-6}$	$n \cdot 10^{-5}$	$< n?$
U	$2.5 \cdot 10^{-4}$	$n \cdot 10^{-5}$	$0.0n$
Hg	$8.3 \cdot 10^{-6}$	$n \cdot 10^{-6} - n \cdot 10^{-5}$	$0.n - n$
Ra	$1 \cdot 10^{-10}$	$n \cdot 10^{-11} - n \cdot 10^{-10}$	$0.n - n$
Sc	$1 \cdot 10^{-3}$	$n \cdot 10^{-6}$	$0.00n$
Zr	0.017	$n \cdot 10^{-4} - n \cdot 10^{-3}$	$0.0n$

*To arrive at the average content of ash we used both the average chemical content of living matter and the direct analyses of plant ashes after A. P. Vinogradov, V. A. Kovda, D. P. Maliuga, A. L. Kovalevsky, B. F. Mitskevich, N. P. Remezov, etc.

since ash components are sorbed by plants from the lithosphere during their life and returned to it or to the hydrosphere upon their death.

As seen in Table 1, the contents of many elements in ash differ considerably from their average abundances in the earth's crust. This is due to the ability of plants to absorb certain elements selectively and to accumulate them. This ability is described as the ratio of the given element's content in the plant ash to its content in the soil on which the plant grew. A grouping of chemical elements on the basis of coefficients of biologic accumulation is given in Table 2.   The data given in these two tables vary one way or another by factors of 10, 100, and even 1000.

Organisms which can accumulate larger amounts of some elements are obviously of practical interest as indicators of those elements. For example, there are numerous groups of marine and land animals which concentrate calcium in their skeletons: mollusks, corals, foraminifera, etc. Upon their death, their skeletons accumulate as sediments which eventually consolidate into lacustrine and marine limestones.

Sponges, radiolaria, and diatoms are typical concentrators of silicon. As a result of their activity, large deposits of silica —

Table 2. Biologic Accumulation Series of Elements

		Coefficients of biologic accumulation					
		$100 \cdot n$	$10 \cdot n$	n	$0.n$	$0.0n$	$0.00n$
Elements of biologic capture	Very strong	P, S, Cl					
	Strong		Ca, K, Mg, Na, Sr, B, Zn, As, Mo, F				
Elements of biologic accumulation	Moderate			Si, Fe, Ba, Rb, Cu, Ge, Ni, Co, Li, Y, Cs, Ra, Se, Hg			
	Weak				Al, Ti, V, Cr, Pb, Sn, U		
	Very weak						Sc, Zr, Nb, Ta, Ru, Rh, Pd, Os, Ir, Pt, Hf, W

tripolite, diatomite, etc. — form in marine and lacustrine eviron-
ments.

Certain microorganisms are capable of concentrating iron or
manganese in their bodies. It has been established that the following elements can concentrate in the bodies of various organisms:
H, Li, Be, B, C, N, O, Na, Mg, Al, Si, P, S, Cl, K, Ca, Ti, V, Cr,
Mn, Fe, Co, Ni, Cu, Zn, Ge(?), Se, Sr, Mo, Ag, I, Au, Pb, Ba, and U.

Growing on the same soil and on the same landscape, different
plants will differ in the composition of their ash. For instance,
cereals always contain more silica than legumes, while calcium is
prominent in the latter. Analyzing the ash of grasses and trees, it
is possible to establish the predominance of manganese in some,
strontium in others, and zirconium in still others. The chemical
composition of organisms is a product of long evolution and is just
as characteristic of certain plants and animals as is their physical
appearance. Conversely, organisms belonging to the same species
will always maintain the same chemical composition in any environment. Thus, legumes contain much calcium in their ash even if
grown on calcium-deficient soil. Copper- or manganese-favoring
plants extract these elements out of soils poor in these elements.
To be sure, geochemical landscapes and aquifers do influence the
composition of such organisms. In regions of ore deposition, where
soils run high in certain metals, plants and animals accumulate
these metals in excess. In this respect, one can speak of a
"biogmic dispersion," which is the prime object of biogeochemical
exploration. Because only some specific organisms in this
biogeochemical halo contain increased amounts of certain metals,
the choice of plants to be sampled should be made with care.

Many regions are noted for their excessive or deficient contents
of some specific elements. Organisms living in such regions react
very definitely toward these chemically abnormal environments by
modifying their size, developing certain diseases, etc. Vinogradov
has referred to such regions as "biogeochemical provinces."

DECOMPOSITION OF ORGANIC MATTER

The air would soon become depleted of its carbon dioxide and
the soil of its mineral salts had biogenic accumulation in the
supergene zone continued without reversals. All life would cease
because of lack of food. This does not happen because nature

operates a counter process and converts living matter back into mineral.

During this process, the energy-rich organic matter is oxidized and is broken down into simple oxides (CO_2, H_2O) and mineral salts. Energy is liberated and goes back to work in the supergene zone.

The destruction of living matter — or its mineralization — takes place even when plants oxidize their tissues while transpiring. However, more living matter is created by photosynthesis than is destroyed by oxidation. In general, plants are accumulators of both organic matter and of energy. Animals, on the other hand, play a larger role in destroying it. Incapable of converting mineral matter into organic, animals destroy organic matter while breathing or during other physiological processes. Microorganisms are the most important creators of mineral matter, since they decompose the plant and animal bodies into CO_2, H_2O, and mineral salts.

Microorganisms thrive in air, water (both the surface and the ground), soils, or the crust of weathering. Their populations in soils may reach several million per gram. Free oxygen is needed for breathing by aerobic bacteria, while anaerobic bacteria need no free oxygen. Sooner or later, microorganisms destroy all proteins, fats, cellulose, chitin, humus, and other organic substances.

The fate of chemical elements liberated during the mineralization of organic matter is unpredictable. Plants obtain C, H, O, and N from CO_2, H_2O, NH_3, and various other simple or complex organics. At the same time SiO_2, Fe_2O_3, and Al_2O_3 are produced. These last interact with each other to form secondary clay minerals. This was surmised by Polynov (1956) who said, "in soils, clays may have a biogenic origin similar to that of humus...."

While living matter forms only at the earth's surface or in the uppermost levels of the sea, where its photosynthesis is dependent on sunlight, its decomposition can take place at considerable depths within the lithosphere. Conditions are favorable for the activity of microorganisms wherever sedimentary or metamorphic rocks contain both water and organic matter. Microbes have been found even in oil production waters at depths of several thousand meters. Deeper portions of the lithosphere are zones of intense chemical activity, where buried organic matter is being destroyed by microbes. The composition of rocks as well as the composition of waters and gases contained in those rocks changes as the processes advance.

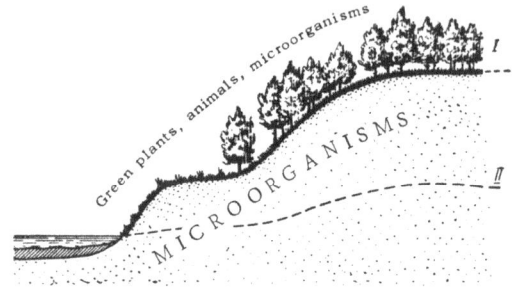

Fig. 2. Principal types of biogenic migration: (I) Living matter predominates over mineralizing matter because of photosynthesis.   (II) Mineralizing matter predominates definitely over living matter; no photosynthesis is feasible.

The decomposition of dead organic matter by microbes produces carbonic and other organic acids, hydrocarbons, etc. Surficial and ground waters become enriched in these products and, in turn, become active in the further destruction of rocks.

Locally, ground water may contain several million bacteria per cubic centimeter. Among them are bacteria that destroy cellulose, remove sulfur or nitrogen, oxidize methane, hydrogen, phenol, naphthalene, etc.   Over geologically long periods of time their activity produces enormous effects.   Huge quantities of organic matter, including guano deposits, are so destroyed.   It can be said that the role of microorganisms is to clear the lithosphere of accumulated dead organic matter.

The so-called iron bacteria concentrate hydrated iron oxides within or around their cells.   Other bacteria concentrate sulfur or other elements.   Many scientists hold that microorganisms played important roles in the formation of ore deposits or iron, manganese, and sulfur.

BIOLOGIC CYCLE OF CHEMICAL ELEMENTS

The mutually opposed processes of biogenic accumulation and mineralization (decomposition) constitute the single biologic cycle of atoms.

While biogenic migration is feasible in all environments of the supergene zone, it may follow different paths.   Thus, organic matter is formed as the result of photosynthesis only on the earth's surface or in the uppermost levels of the sea.   Mineralization of that matter may occur in the same environment, but is hardly ever completed before getting buried.

No photosynthesis is feasible in marine or lacustrine muds or in ground water. The formation of any organic matter — bodies of bacteria — occurs there at the expense of dead matter and not mineral (CO_2, H_2O, etc.). In contrast to surface environments, decomposition of organic matter predominates here over its creation. Here are decomposed those organic substances which were formed in the upper parts of the supergene zone. Obviously, the biologic cycle of atoms also continues here, although its intensity and its path are different. Two main groups of environments can be distinguished on this basis (Fig. 2):

1. Areas of dry land and upper portions of the sea where living matter forms at the expense of mineral matter and eventually reverts to it.

2. Weathered crust, water-bearing strata and deeper levels of the sea where photosynthesis is precluded by lack of light. Here organic matter is mostly decomposed by microorganisms. New masses of living matter (bacteria, deep sea fish, etc.) are formed at the expense of the dead organic matter and not mineral matter. The balance of the biologic cycle is negative. Consequently, lower horizons of the supergene zone are regions of decomposition of organic matter which accumulated at the earth's surface.

While such physical processes as the cycling of water, heating of the earth's surface, or movement of air masses are motivated by solar energy, chemical processes operating within the supergene zone are more complex.

The principal source of chemical energy there is again solar energy but its action is indirect, having been channeled hrough the biologic cycle of atoms. This energy is liberated in a form capable of doing the chemical work of mineralizing organic substances. It follows that the principal geochemical features of the supergene zone result from the biologic cycling of atoms, i.e., they are conditioned by living matter. This geochemical premise was first formulated by Vernadskii and later restated by Polynov.

Biologic cycles are not completely reversible. The supergene zone does not return to its original state upon completion of a cycle, but progresses forward. For example, in the forest zone sapropel accumulates on lake bottoms. These lake bottoms rise as lake basins become filled with it. In time, lakes become bogs.

Biogenic migrations obey complicated biologic laws in addition to laws of physics and chemistry. The behavior of elements during biogenesis, however, is determined by such characteristics as ionic radii, valence, polarization, and electronegativity.

Elements accumulated in living matter possess different chemical properties, and even related elements behave differently during biogenesis.

Most elements are held in organic compounds by covalent and other nonionic bonds, while in inorganic compounds they are held mostly by ionic bonds. Therefore, the behavior of such metals as Ca, Mg, Na, and K should be different when included in living matter.

Elements participating in biogenesis include some typical atmophiles (N), lithophiles (Si, K, Mg, Ca, Na, etc.), chalcophiles (Cu, Zn, etc.), and siderphiles (Fe). Associations of such widely different elements cannot be explained satisfactorily by the different solubilities of their inorganic compounds. Some specific bio-chemical relationships should be instrumental. The biogenic migration of elements is not always governed by their chemical properties. For example, huge quantities of bird guano are ac-cumulating on some islands along the coast of Chile and Peru which are the nesting grounds of birds. The migration of phosphorus contained in that guano is controlled by laws of bird migration and not by its chemical properties.

LIVING MATTER IN GEOLOGIC HISTORY

One can speak of diverse aspects of the geochemical activity of living organisms. Above, we considered the role of living matter restricted to some definite portion of the supergene zone. Inte-grating the work of living matter over the entire geologic time, we obtain truly staggering results.

One of the products of the geologic activity of living matter is the change of atmospheric composition. During a span of a few billion years, plants have cleared from the atmosphere practically all carbon dioxide, substituting oxygen for it. Being biogenic in origin, atmospheric oxygen is a product of photosynthesis. No other reaction of oxygen evolution could have been common enough on the earth. Removal of carbon dioxide from the air has brought about enrichment of the lithosphere in carbon through the formation of limestones, dolomites, eoal, or oil. The organic carbon of sedimentary rocks is far from being chemically inert. It is food for microorganisms, and in the process of thert more lism ground waters become enriched in CO_2 H_2S, etc. As geologic history progressed, more and more organic matter became trapped in sedimentary rocks. The geochemical activity of ground water was increasing at the same pace.

All the more important components of surface waters — e.g., CO_2, HCO_3-, Ca^{2+} — are products of the chemical activity of living matter.

This has led to the concept of the biosphere, which was first introduced by Zeuss and later developed by Vernadskii.

The biosphere is defined as that portion of the earth's crust which is populated by organisms and is characterized by geologic activity of all forms of living matter — plants, animals, and microorganisms.

In the biosphere, chemical processes go on in the presence of organisms or among their physicochemical products (CO_2, O_2, H_2S, etc.). Since microorganisms have been found in deep-seated oil production waters, the biosphere extends down into the lithosphere for thousands of meters in some regions. It is thinnest in regions covered by recent volcanic lavas.

According to Pokrovskii (1961), the lower limit of the biosphere should lie along the 100°C isotherm, this temperature being critical for bacterial growth. This limit lies at depths of 10,000 to 15,000 m on the Baltic and Ukrainian shields of the European USSR. In young synclines elsewhere it rises to within 2000-1500 m from the surface. On the Russian Platform it fluctuates between 5500 and 3300 m. Since the world's oceans are populated at all depths, the entire hydrosphere plus oceanic and marine bottoms should be included in the biosphere. Similarly, the troposphere should also be included. This constitutes Vernadskii's concept of the biosphere* and is illustrated on Fig. 3.

A very practical conclusion may be drawn on the basis of the biosphere concept and the role of organisms: Supergene processes should be irreversible during geologic history, since living matter

---

*A serious error is incurred occasionally in the literature and particularly in textbooks where the biosphere is visualized as a film of life enclosing the earth's crust.

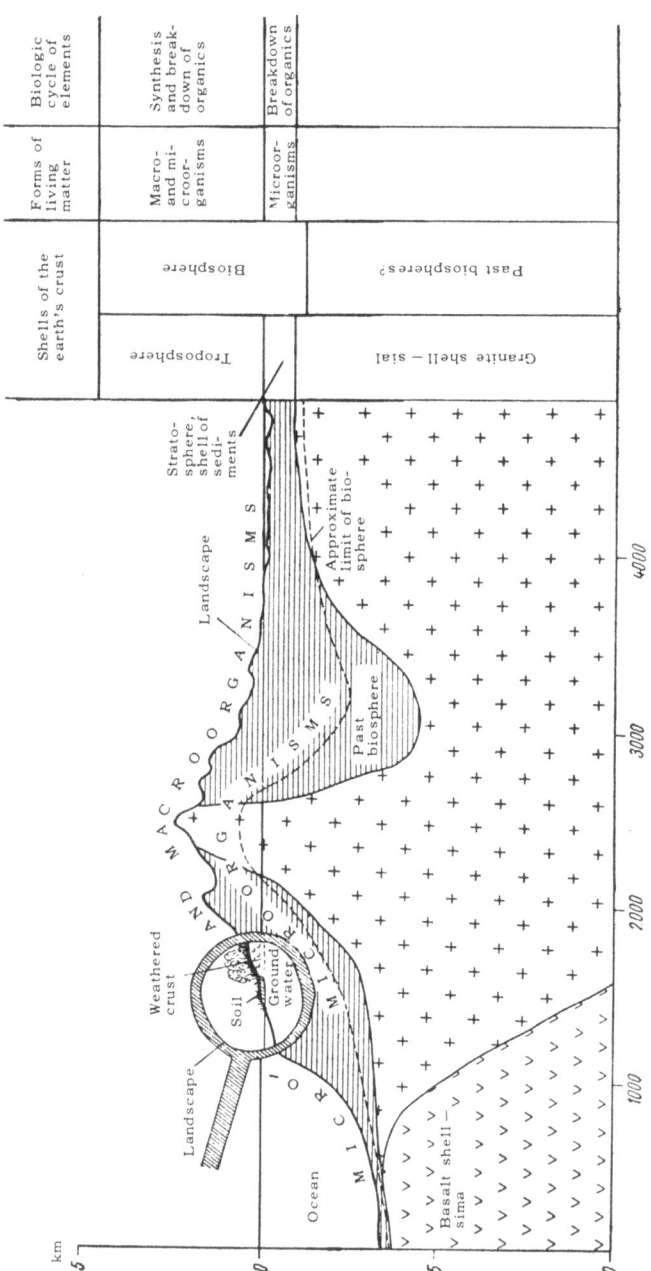

Fig. 3. Structure of the biosphere.

has been evolving progressively and dead matter has been accumulating in rocks at an ever increasing rate.

The composition of the atmosphere must have been different during the Paleozoic era and especially during the Precambrian period. It must have contained more carbon dioxide and less oxygen. During still earlier time (Archaeozoic?) the atmosphere probably lacked oxygen and thus differed radically from the present one.

The migrational abilities of iron and manganese — probably in bivalent state — were accentuated, since there was no oxidation. Compositions of the hydrosphere and of soils must have been different. In fact, sedimentation must have been changing ever since. During the former eras there formed sedimentary rocks and ores which at the present form in minor amounts or do not form at all. For example, sedimentary iron ores of the type of ferruginous quartzites precipitated only in the Precambrian period. In fact, throughout the entire history of the earth, iron sedimentation has been most typical of the Precambrian period. The greatest reserves of iron ores were deposited then. On the other hand, the formation of coal and oil took place at much later stages of the earth's history. Salt deposits are also of the post-Proterozoic era. Problems of sedimentation and ore formation throughout the earth's history have been investigated by Strakhov (1960-1963).

Extrusive igneous rocks have been weathered and denuded repeatedly throughout the earth's history. As a result sediments and sedimentary rocks were formed. Sinking to great depths in synclinal zones these reverted back to magmatics. In this way the idea of a great turnover of matter is built. It is possible that magmatic phenomena are closely related to exogenous processes. To understand the endogene mineralization, it should be considered together with the sedimentation that preceded it, climatic zoning, and all other exogenous factors.

Chapter 2

Geochemical Parameters of Elements
Which Condition Their Physical
and Chemical Migration

Physicochemical migration is not very different from the biogenic, since such phenomena as acidity–basicity, oxidation–reduction, or the gaseous content of natural waters are controlled by organisms. However, should some element become dissolved in the form of an ion, an undissociated molecule, or a colloidal miscela, its behavior will be governed by the laws of physics and chemistry.

The physicochemical migration of an element depends on its ability to react with other elements, to change its valence, to form soluble or gaseous compounds, to be sorbed in colloids, etc. All such properties depend on the structure of its atom and, particularly, on the structure of its valence shell.

The particular types of chemical bonds, electronegativity, ionic or atomic radii, and valence conditioning factors are referred to as the "geochemical parameters" of the elements.

CHEMICAL BONDS AND CRYSTAL LATTICES OF MINERALS

Bonds between atoms regulate the structure of crystal lattices and, consequently, the properties of minerals.

I o n i c (heterovalent, heteropolar) bonding takes place when one of the interacting atoms loses one or more electrons to another. As a result, the first atom acquires a positive charge and becomes a cation, while the atom which received those electrons becomes charged negatively and is then known as an anion. Ions with opposite

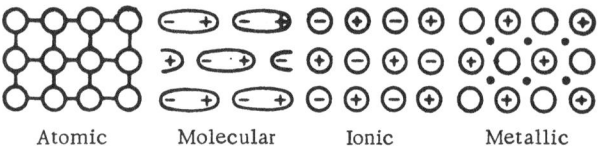

Atomic   Molecular   Ionic   Metallic

Fig. 4. Principal structures of solids.

charges are attracted to each other and form crystals. Only those ions are bonded ionically which differ radically from each other.

C o v a l e n t (atomic, homopolar) bonding is common between atoms which resemble each other.   There is no transfer of electrons, but valence electrons of both form couples which are shared by both atoms. Should these atoms be identical, the electron couple is shared equally by each atom. This particular type is known as a nonpolar bond. Examples are gas molecules such as H_2, F_2, etc.

However, the most common interactions under conditions in the earth's crust are between unlike atoms. In such cases the electron couple is more attracted to one atom without being actually torn out of the other. Such bonding is intermediate between ionic and non-polar and is known as polar (e.g., H_2O or CO_2).

M e t a l l i c bonding resembles both the ionic and the covalent. In metals, strongly charged cations are held together with electrons which move from one anion to another. As it were, they form an electronic gas and thus condition conductivity and other electric properties.

Some substances are composed of molecules held together by weak intermolecular bonds while individual ions within the molecule are bound with strong covalent bonds.

Different types of chemical bonding predetermine differences in structures of crystal lattices, there being distinguished four main types (Fig. 4):

1. Molecular
2. Coordinated ionic
3. Coordinated atomic and metallic
4. Intermediate between molecular and coordinated

Individual molecules are situated at vertices of crystal lattices. Bonding between individual atoms within a molecule is covalent and polar, while molecules themselves are held together by relatively weak intermolecular forces. A comparatively small amount of energy would be needed to destroy such a lattice. For this reason, substances with molecular lattices go into the liquid or gas state relatively easily. This is the lattice type possessed by CO_2, HCl, H_2O, H_2S, SnF_4, UF_6, WF_4, and by most organic compounds, including hydrocarbons.

All such compounds have relatively low melting and boiling points, which account for their important roles during pneumatolytic and hydrothermal processes.

Individual ions located at the vertices of coordinated ionic lattices are bound together with firm electrostatic (ionic) bonds which are considerably stronger than the intermolecular forces. Therefore, such lattices characterize stable compounds, since more energy would be required for their disruption. The melting and boiling temperatures of substances with such lattices are usually high. Many such substances are readily soluble in water, where they ionize. This explains their importance in aquatic migration. Chloralkalies like NaCl are typical examples of substances with such lattices.

In coordinated atomic lattices, atoms occupying vertices are firmly bonded together with covalent bonds. Such substances possess high melting and boiling points. Diamond is an example. Crystal lattices of metals resemble somewhat the ionic lattices but are closer to the atomic. Weakly bound electrons are characteristic of metal lattices.

Crystal lattices of the intermediate type are most common among naturally forming compounds. They display features similar to both the molecular and the coordinated lattices.

Electronegativity measures the relationship of the atoms to the electrons in a given compound. It is measured in calories, electron-volts, or in some arbitrary units (Table 3). Halogens display the greatest electronegativity and alkali metals the least. Elements with different electronegativities form ionic bonds and those with similar electronegativities form covalent (polar or nonpolar) bonds. Thus, alkali halides display typical ionic lattices, electronegativities being 0.9 for Na and 3.0 for Cl. Heavy metal sulfides are characteristically bound with intermediate covalent bonds, electronegativities being 2.5 for sulfur, 2.0 for copper, 1.7 for iron, 1.8 for nickel, 1.6 for lead, etc.

Crystal lattices of the first type ionize in aqueous solutions, while lattices of the second type do not fall apart.

Large identical values of electronegativity for the atoms in the lattice of diamond account for the covalent, nonpolar character of

Table 3. Electronegativity

Period	Ia	IIa	IIIa	IVa	Va	VIa	VIIa	VIIIa	VIIIa	VIIIa	Ib	IIb	IIIb	IVb	Vb	VIb	VIIb
1																	H 2.15
2	Li 0.95	Be 1.5											B 2.0	C 2.5	N 3	O 3.5	F 3.95
3	Na 0.9	Mg 1.2											Al 1.5	Si 1.8	P 2.1	S 2.5	Cl 3.0
4	K 0.8	Ca 1.0	Sc 1.3	Ti 1.6	V 3+1.4 4+1.7 5+1.9	Cr 2+1.4 3+1.5 4+2.2	Mn 2+1.4 3+1.2 7+2.5	Fe 2+1.7 3+1.8	Co 1.7	Ni 1.8	Cu 1+1.8 2+2.0	Zn 1.5	Ga 1.5	Ge 1.8	As 2.0	Se 2.4	Br 2.8
5	Rb 0.8	Sr 1.0	Y 1.2	Zr 1.5	Nb 1.7	Mo 4+1.6 5+2.1 6+2.1	Tc 5+1.9 7+2.4	Ru 2.0	Rh 2.1	Pd 2.0	Ag 1.8	Cd 1.5	In 1.5	Sn 2+1.7 4+1.8	Sb 3+1.8 5+2.1	Te 2.1	I 2.55
6	Cs 0.75	Ba 0.9	La⇄ 1.1	Hf 1.4	Ta 3+1.3 4+1.5 5+1.7	W 4+1.6 6+2.0	Re 5+1.8 7+2.2	Os 2	Ir 2.1	Pt 2.1	Au 2.3	Hg 1.8	Tl 1+1.5 3+1.9	Pb 2+1.6 4+1.8	Bi 1.8	Po 2.0	At 2.2
7	Fr 0.7	Ra 0.9	Ac⇄ 1.1														

Lanthanides ⇄

IVa	Va	VIa	VIIa	VIIIa	VIIIa	VIIIa	Ib	IIb	IIIb	IVb	Vb	VIb	VIIb
Ce 1.1	Pr 1.1	Nd 1.2	Pm 1.2	Sm 1.2	Eu 1.1	Gd 1.2	Tb 1.2	Dy 1.2	Ho 1.2	Er 1.2	Tu 1.2	Yb 1.1	Lu 1.2

Actinides ⇄

IVa	Va	VIa	VIIa	VIIIa	VIIIa	VIIIa	Ib	IIb	IIIb	IVb	Vb
Th 2+1.0 3+1.3 4+1.4	Pa 3+1.3 4+1.5 5+1.7	U 4+1.4 6+1.9	Np 1.1	Pu 1.3	Am 1.3	Cm 1.3	Bk 1.3	Cf 1.3	E	Fm	Mv

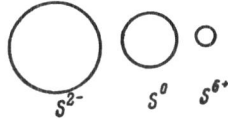

Fig. 5. Effect of valence upon atomic dimensions.

the chemical bonding which precludes ionization. Small though comparable values of electronegativity favor the formation of metallic lattices of metals and alloys.

VALENCE

The greater the valence of metals, the less soluble are their compounds and, accordingly, the lower is their migrational ability. As a rule, compounds of monovalent alkali metals are most soluble ($NaCl$, Na_2SO_4, etc.). Bivalent metals of the alkaline earth group are less soluble. This is obvious if we compare solubilities of easily soluble Na_2SO_4 and those of difficultly soluble $CaSO_4$ or $CaCO_3$. Even less soluble are compounds of the trivalent metals Al and Fe. A similar rule holds also for anions: The PO_4^{3-} ion forms less soluble compounds than do SO_4^{2-} or Cl^-.* In reality this rule is often modified by polarization and other phenomena. Such influences account, for example, for the low solubility of AgCl.†

Heterovalent ions of the same chemical element display different properties and different migrational abilities. In fact, they behave like entirely different elements. For instance, iron behaves like aluminum or chromium when within the crust of weathering, while bivalent iron resembles calcium. The trivalent and pentavalent ions of vanadium, the bivalent and hexavalent ions of sulfur, and the tetravalent and hexavalent ions of uranium have entirely different behaviors.

Not all possible valences are met with during epigenetic processes in the supergene zone. For example, compounds of heptavalent chlorine or manganese, such as $KMnO_4$ or $HClO_4$, have not been found in nature although they are synthesized in the laboratory.

*Such rules hold only for minerals with ionic lattices.

†This has been explained by Fersman in his geoenergetic theory (1937).

Table 4. Ionic Radii

Period	Subgroup							
	Ia	IIa	IIIa	IVa	Va	VIa	VIIa	VIIIa
1								
2	Li 1^+ 0.68	Be 2^+ 0.34						
3	Na 1^+ 0.98	Mg 2^+ 0.74						
4	K 1^+ 1.33	Ca 2^+ 1.04	Sc 3^+ 0.83	Ti 2^+ 0.78 3^+ 0.69 4^+ 0.64	V 2^+ 0.72 3^+ 0.67 4^+ 0.61 5^+ 0.4	Cr 2^+ 0.83 3^+ 0.64 6^+ 0.35	Mn 2^+ 0.91 3^+ 0.70 4^+ 0.52 7^+ (0.46)	Fe 2^+ 0.80 3^+ 0.67 Co 2^+ 0.78 3^+ 0.64
5	Rb 1^+ 1.49	Sr 2^+ 1.20	Y 3^+ 0.97	Zr 4^+ 0.82	Nb 4^+ 0.67 5^+ 0.66	Mo 4^+ 0.68 6^+ 0.65	Tc	Ru 4^+ 0.62 Rh 5 3^+ 0.75 4^+ 0.6
6	Cs 1^+ 1.65	Ba 2^+ 1.38	La 3^+ 1.04 4^+ 0.90	Hf 4^+ 0.82	Ta 5^+ (0.66)	W 4^+ 0.68 6^+ 0.65	Re 6^+ 0.52	Os 4^+ 0.65 Ir 4^+ 0.65
7	Fr	Ra 2^+ 1.44	Ac 3^+ 1.11					
Lanthanides				Ce 3^+ 1.02 4^+ 0.88	Pr 3^+ 1.00	Nd 3^+ 0.99	Pm 3^+ (0.98)	Sm 3^+ 0.97 Eu 3^+ 0.97
Actinides				Th 3^+ 1.08 4^+ 0.95	Pa 3^+ 1.06 4^+ 0.91	U 3^+ 1.04 4^+ 0.89	Np 3^+ 1.02 4^+ 0.88	Pu 3^+ 1.01 4^+ 0.86 Am 3^+ 1.00 4^+ 0.85

Note: 1. Calculated radii are given in parentheses. 2. Atomic radii are given for noble (inert) gases.

Subgroup							
Ib	IIb	IIIb	IVb	Vb	VIb	VIIb	VIIIb
						H 1^- 1.36 1^+ 0.00	He 0 1.22
		B 3^+ (0.20)	C 4^+ 0.2 4^+ (0.15) 4^- (2.60)	N 3^+ 5^+ 0.15 3^- 1.48	O 2^- 1.36	F 1^- 1.33	Ne 0 1.60
		Al 3^+ 0.57	Si 4^+ 0.39	P 3^+ 5^+ 0.35 3^- 1.86	S 2^- 1.82 6^+ (0.29)	Cl 1^- 1.81 7^+ (0.26)	Ar 0 1.92
Ni 2^+ 0.74	Cu 1^+ 0.98 2^+ 0.80	Zn 2^+ 0.83	Ga 3^+ 0.62	Ge 2^+ 0.65 4^+ 0.44	As 3^+ 0.69 5^+ (0.47) 3^- 1.91	Se 2^- 1.93 4^+ 0.69 6^+ 0.35	Br 1^- 1.96 7^+ (0.39)	Kr 0 1.98
Pd 4^+ 0.64	Ag 1^+ 1.13	Cd 2^+ 0.99	In 3^+ 0.92	Sn 2^+ 1.02 4^+ 0.67	Sb 3^+ 0.90 5^+ 0.62 3^- 2.08	Te 2^- 2.11 4^+ 0.89 6^+ (0.56)	J 1^- 2.20 7^+ (0.50)	Xe 0 2.18
Pt 4^+ 0.64	Au 1^+ (1.37)	Hg 2^+ 1.12	Tl 1^+ 1.49 3^+ 1.05	Pb 2^+ 1.26 4^+ 0.76	Bi 3^+ 1.20 5^+ (0.74) 3^- 2.13	Po	At	Rn

Gd 3^+ 0.94	Tb 3^+ 0.89	Dy 3^+ 0.88	Ho 3^+ 0.86	Er 3^+ 0.85	Tu 3^+ 0.85	Yb 3^+ 0.81	Lu 3^+ 0.80

Cm	Bk	Cf	E	Fm	Mv		

Table 5. Atomic Radii

Period	Ia	IIa	IIIa	IVa	Va	VIa	VIIa	VIIIa			Ib	IIb	IIIb	IVb	Vb	VIb	VIIb	VIIIb
1																	H 0.46	He 1.22
2	Li 1.55	Be 1.13											B 0.91	C 0.77	N 0.71	O	F	Ne 1.60
3	Na 1.89	Mg 1.60											Al 1.43	Si 1.34	P 1.3	S	Cl	Ar 1.92
4	K 2.36	Ca 1.97	Sc 1.64	Ti 1.46	V 1.34	Cr 1.27	Mn 1.30	Fe 1.26	Co 1.25	Ni 1.24	Cu 1.28	Zn 1.39	Ga 1.39	Ge 1.39	As 1.48	Se 1.6	Br	Kr 1.98
5	Rb 2.48	Sr 2.15	Y 1.81	Zr 1.60	Nb 1.45	Mo 1.39	Tc 1.36	Ru 1.34	Rh 1.34	Pd 1.37	Ag 1.44	Cd 1.56	In 1.66	Sn 1.58	Sb 1.61	Te 1.7	J	Xe 2.18
6	Cs 2.68	Ba 2.21	La 1.87	Hf 1.59	Ta 1.46	W 1.40	Re 1.37	Os 1.35	Ir 1.35	Pt 1.38	Au 1.44	Hg 1.60	Tl 1.71	Pb 1.75	Bi 1.82	Po	At	Rn
7	Fr 2.80	Ra 2.35	Ac 2,03															

Subgroup

Lanthanides:

Ce 1.83	Pr 1.82	Nd 1.82	Pm	Sm 1.81	Eu 2.02	Gd 1.79	Tb 1.77	Dy 1.77	Ho 1.76	Er 1.75	Tu 1.74	Yb 1.93	Lu 1.74

Actinides:

Th 1.80	Pa 1.62	U 1.53	Np 1.50	Pu 1.62	Am	Cm	Bk	Cf	E	Fm	Mv

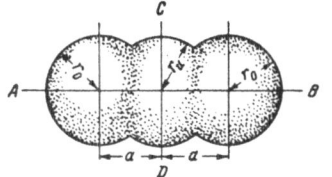

Fig. 6. Shape and size of uranyl ion (after Lipilina). AB — axis of UO_2; ionic radius of U — 1.40 Å; ionic radius of O — 1.32 Å.

DIMENSIONS OF IONS AND ATOMS

Not the least important chemical characteristic of an element is its atomic or ionic radius.

Introduced by Goldschmidt, the concept has been useful in the development of this science. An entirely new dimension has been added — the size of the atom or its ion. Provisionally, Goldschmidt had assumed that atoms and ions have spherical shapes which are not compressible. Another condition was that apparent and not true radii should be used, i.e., the radii of spheres of influences and not the radii of actual atoms or ions.

Ionic radii are measured in Angstrom units, Å, the sizes fluctuating about 1 Å, which is equal to $1 \cdot 10^{-8}$ cm. It has been established that ions of different valence will have different sizes for the same element. For example, the radius of bivalent sulfur is 1.90 Å and that of hexavalent sulfur is 0.29 Å (Fig. 5).

As a rule, anionic radii are larger than cationic radii. The structure of the crystal lattice, and especially the coordination number, influence the size of ions. One considers ionic radii in ionic lattices and covalent radii in atomic or metallic lattices (Tables 4 and 5). Atomic radii are given for a coordination number of 6.

Ionic radii can be calculated even for complex ions, such as SO_4^{2-}, CO_3^{2-}, PO_4^{3-}, UO_2^{2+}. Their size is larger and the shapes may not be spherical. In such cases one can speak of several radii (Fig. 6). For a given element, ionic radius decreases as the valence increases: Fe^{2+} — 0.80, Fe^{3+} — 0.67.

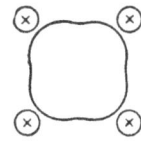

Fig. 7. Polarization of an ion within a crystal.

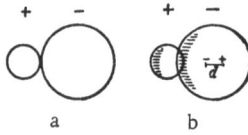

Fig. 8. Interaction between two ions: (a) without considering polarization; (b) considering this polarization.

There is another reason for the conditional nature of the premise that ions are incompressible spheres. In crystals, each ion is situated within the electric fields of other ions. Because of this, the center of gravity of the charges is displaced and the ion is no longer spherical but becomes "polarized" (Figs. 7 and 8). The larger the ionic radius, the smaller is the ionic charge and the easier it is to polarize the ion. This is particularly well marked on anions. Polarization is particularly strong in the case of bivalent sulfur. Cations are capable of polarizing the neighboring anions, this being particularly true of ions possessing an 18-electron shell, such as copper or silver.

Various phenomena like isomorphism become understandable when one considers the dimensions of atoms and ions.

ISOMORPHISM

Isomorphism is the ability of an element to substitute for another, provided that the atomic (or ionic) radii and the polarization of the two are comparable and electric charges are balanced (Fersman, 1939).

Goldschmidt likened the distribution of rare elements during magmatic differentiation to the passage of a powder through a nest of sieves of different size. A number of rare elements are dragged down together with more common ones and are trapped in the crystal lattices of rock-forming minerals. For example, bivalent nickel is dragged down together with bivalent iron or magnesium since the radii of the three are comparable and electric charges are the same.

Many peculiarities in the distribution of elements in rocks, ores, and minerals, and particularly the so-called mineral assemblages can now be explained by isomorphism. To mention a few, there are the associations of gold and silver and of platinum and platinum metals, the occurrence of cobalt and nickel in silicates of iron and magnesium, and the presence of lead and barium in potassium minerals. A theoretical analysis of isomorphism facilitates predictions of occurrence of some disperse elements and thus contributes to a better planning of exploration.

Isomorphism is mainly responsible for formation of syngenetic haloes of dispersion about hypogene ore deposits.

At elevated temperatures near those of mineral melting points, isomorphous substitutions are possible even if the radii of substituting and substituted ions differ by as much as 20 to 30%. At the low temperatures prevailing at the earth's surface, these are achieved only if radii differ by not more than 10 to 15%. Thus, potassium and sodium substitute for each other in feldspars of magmatic origin but not in minerals formed within the supergene zone. In general, isomorphous substitutions decrease within the supergene zone.

Oxidation reactions which are so typical of the supergene zone alter the valence of elements and, consequently, the radii of their ions. This also diminishes chances for isomorphous substitution. For example, magnesium and bivalent cobalt or iron are isomorphous in olivines and other magmatic minerals. During weathering, iron and cobalt become trivalent and their radii diminish so that no substitution for magnesium is possible. As a result, iron and cobalt drop out of lattices of magnesian minerals. This phenomenon is known as autolysis or self-purification (Fersman). Containing fewer isomorphous admixtures, supergene minerals are purer than hypogene minerals. Therefore, isomorphous substitutions are less significant to the formation of secondary haloes of dispersion or primary haloes in sedimentary rocks of the supergene zone than for hypogene environments. Instead, such factors as the sorption of ions by colloids or by living matter acquire importance. The variation of isomorphous substitution in different vertical zones of the earth's crust has been portrayed in Vernadskii's empirical series of isomorphism.

The concept of atomic and ionic radii has been very useful in the development of geochemistry, since it explained much as to the distribution of rock-forming elements. There are still some unaccounted phenomena. For example, a few sodium minerals contain copper and a few copper minerals contain sodium, although the ionic radii of both elements are identical — 0.98.

Recent investigations have shown that the concept of ionic radii is best applicable to chemical elements of the I, II, III, and VII groups of the periodic system of elements. Elements of other

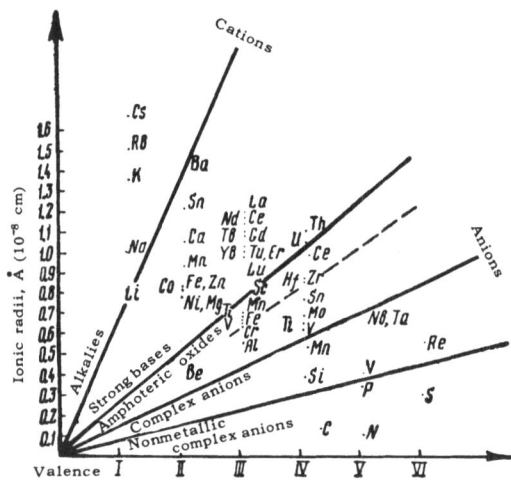

Fig. 9. Acidic and basic properties of elements as functions of valence and ionic radius
(stated by Goldschmidt and revised by Shcherbina).

groups do not normally enter crystal lattices in the form of typical
ions. In those cases it would be better to speak of covalent radii.
For this reason, the structure of some minerals like spinels does
not correspond to the above–discussed views.*

IONIC POTENTIAL

The ratio of the valence charge on a cation to its radius is
referred to in geochemistry as J, its ionic potential† (Fig. 9). It
varies between 0.5 and 4.7 for elements like K, Na, Ca, Mg, Sr, Ba,
Li, La, Si, Al, Zr, Hf, etc., which have cations with completed
eight-electron outer shells. Such a structure conditions their
being bases. The smaller J, the more alkaline they behave when
in solution: 1.02 for Na, 1.92 for Ca. At J values between 4.7 and
8.6, elements form compounds which are amphoteric when in
aqueous solution. Finally, at J above 8.6, complex anions form
in aqueous solutions (NO_3^-, PO_4^{3-}, etc.).

*For greater details see the work by Lebedev (1957).

†This term is unfortunate, since it denotes no potential in the physical sense of the word
  and may be confused with the ionization potential.

For Cu, Zn, Pb, and other elements with completed 18-electron shells, the relationships are different. For J values under 2, elements are basic. They are amphoteric for values between 2 and 6 and acidic for values above 6.

Thus, it is possible to predict the acidity–basicity of an element in natural solution if its ionic radius, valence, and structure are known.

SOLUBILITY

The solubility of rocks and minerals in natural waters depends upon the ionic radii, valence, polarization, types of chemical bonds, and many other properties of the elements involved. In turn, solubility is closely controlled by such external factors as temperature, pressure, concentration, and Eh and pH of waters.

Ionically bonded minerals are more soluble in water than minerals held together with covalent bonds. For example, sulfates consisting of cations and anions are more soluble than sulfides with their covalent bonds. Silver fluoride possesses an ionic lattice and is quite soluble, while silver chloride with its covalent bonds is poorly soluble.

Solubility increases directly as the ionic radius and inversely with the valence for minerals held with ionic bonds. Thus, Na_3PO_4,

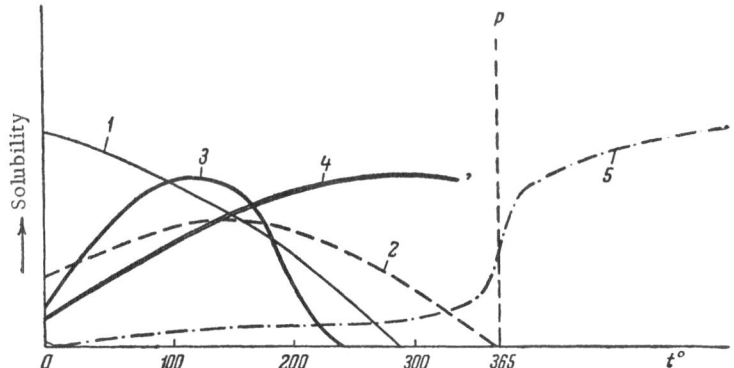

Fig. 10. Different types of solubilities of salts at constant pressure (about 1–5 atm) and rising temperature (after Fersman, 1934). (1) $CaSO_4$ type; (2) NaCl type; (3) sulfate type; (4) type of some halides; (5) silicates (after Niggli). P is the critical point of water.

Table 6. Solubilities of Chemical Compounds in Distilled Water at 20° C and 1 atm

Compound	Solu-bility	Compound	Solu-bility	Compound	Solu-bility
		Easily soluble − over 2g/liter			
$SbCl_3$	9315	$NaCl$	360	Na_2SO_4	194
$ZnCl_2$	3680	$BaCl_2$	357	$SnSO_4$	190
UO_2Cl_2	3200	KCl	340	$AlK(SO_4)_2 \cdot 12H_2O$	114
$SnCl_2$	2698 (15)	$HgCl_2$	66	K_2SO_4	111.1
$CsCl$	1865	$PbCl_2$	9.9	$Ce_2(SO_4)_3$	101
$CdCl_2$	1350	$TlCl$	3.0	Tl_2SO_4	49
$FeCl_3$	919	$Fe_2(SO_4)_3 \cdot 9H_2O$	4400	$Th(SO_4)_2 \cdot 9H_2O$	16
$RbCl$	912	Cs_2SO_4	1786	Rb_2CO_3	4500
$LiCl$	785	$CdSO_4$	770	Cs_2CO_3	2607
$CaCl_2$	745	$MnSO_4$	640	K_2CO_3	1117
$MnCl_2$	740	$ZnSO_4$	544	Na_2CO_3	215
$CuCl_2$	730	$BeSO_4 \cdot 4H_2O$	425	Tl_2CO_3	42
$AuCl_3$	680			$AgNO_3$	2180
$FeCl_2$	644	$NiSO_4$	380	Na_2WO_4	724
$NiCl_2$	642	$MgSO_4$	360	NaF	42
$MgCl_2$	545	$CoSO_4$	344	$Na_2B_4O_7$	26
$SrCl_2$	530	Li_2SO_4	342	LiF	3.0
$CoCl_2$	530	$FeSO_4 \cdot 7H_2O$	265		
NH_4Cl	375	$CuSO_4$	207		
		Difficultly soluble − 2 to 0.1 g/liter			
$CaSO_4$	2.0	$ZnCO_3$	0.2	PbF_2	0.65
Hg_2SO_4	0.6	$Fe(HCO_3)_2$	0.43	$Ca(HPO_4)_2 \cdot 2H_2O$	0.2
$SrSO_4$	0.114			SiO_2	0.16
$CaCO_3 \cdot MgCO_3$	0.21				
		Very difficultly soluble − 0.1 to 0.001 g/liter			
$PbSO_4$	0.041	$Mn(OH)_2$	$1.7 \cdot 10^{-2}$	Al_2O_3	$1 \cdot 10^{-3}$
$BaSO_4$	$2.3 \cdot 10^{-3}$	$Mg(OH)_2$	$9 \cdot 10^{-3}$	$AgBr$	$1 \cdot 10^{-4}$
Hg_2Cl_2	$2 \cdot 10^{-3}$	$Zn(OH)_2$	$6.1 \cdot 10^{-3}$	CaF_2	0.016
$AgCl$	$1.5 \cdot 10^{-3}$	$Ni(OH)_2$	$5.6 \cdot 10^{-3}$		
$MnCO_3$	0.065	$Co(OH)_2$	$5.5 \cdot 10^{-3}$		
Ag_2CO_3	0.032	$Al(OH)_3$	$1 \cdot 10^{-3}$		
$FeCO_3$	$1-3 \cdot 10^{-2}$	$Fe(OH)_2$	$7.5 \cdot 10^{-4}$		
$BaCO_3$	0.022	$Fe(OH)_3$	$5 \cdot 10^{-4}$		
$CaCO_3$	$1.6 \cdot 10^{-2}$	MgO	$6.2 \cdot 10^{-3}$		
$PbCO_3$	$1.1 \cdot 10^{-3}$	HgO	0.05		
$SrCO_3$	0.011	ZnO	$1.6 \cdot 10^{-3}$		
		Practically insoluble − less than 0.0001 g/liter			
FeS_2	−	Ag_2S	−		
PbS	−	TiO_2	−		
ZnS	−	AgJ	$3 \cdot 10^{-6}$		
CuS	−	$PbCrO_4$	$4.3 \cdot 10^{-5}$		
Cu_2S	−	Cu	−		
HgS	−	Au	−		

Na_2SO_4, and Na_2CO_3 are respectively more soluble than $Ca_3(PO_4)_2$, $CaSO_4$, or $CaCO_3$. Similarly, the fluorides of Mg, Ca, Sr, Ba, Al, Sc, Y, and La are less soluble than the fluorides of Na, K, Rb, and Cs. Possessing a small cation, lithium is poorly soluble when in fluoride form.*

The solubility of solid minerals and gases varies with temperature, the relationships being direct for some (SiO_2, NaCl) or inverse for others (O_2, CO_2, and other gases) (Fig. 10).

As temperature rises, the migration of silica in hydrothermal solution becomes more energetic and produces extensive silications of rocks as well as quartz veins. At the low temperatures prevailing at the earth's surface, these phenomena are less impressive. At the very low temperatures observable in permafrost regions, the solubility of O_2 or CO_2 in water increases sharply in comparison to hot climates.

Pressure also affects solubilities. For gases and most solid minerals, solubility increases directly as pressure. This conditions the instability of rocks like limestone when deeply buried.

Solubilities of minerals are given in Table 6 for 20°C and 1 atm. It may be readily seen that fluctuations are considerable.

In chemistry, compounds are said to be insoluble if they are soluble less than $1 \cdot 10^{-3}$ to $1 \cdot 10^{-4}$ g/liter of distilled water. Such figures may mean either low or high solubility in geochemistry, since it all depends on the relative abundance of the given element. For example, such a content would be very low for iron, which comprises 4.65% of the lithosphere. The same figure of $1 \cdot 10^{-4}$ g per liter would be very high for uranium, which has lithospheric abundance of only $3 \cdot 10^{-4}$%. The presence of $n \cdot 10^{-6}$ g/liter of nickel, cobalt, copper, or molybdenum in water does not mean their low solubility but their low abundances. A low content of common elements in natural water indicates low solubilities of their minerals, while low content of rare† elements is conditioned by both low solubility and low abundance. Very low contents of such elements as lithium, rubidium, cesium, selenium, bromine, or iodine are due to their low abundances. Individual solubilities of their various compounds are generally high.

*It was shown by B. K. Yatsimirsky that an ionic compound becomes stabler as differences in the size of its constituent ions decrease. Thus, $BaSO_4$ is less soluble than $BaCrO_4$ and $MgSO_4$, the ionic radii of the involved ions being: Ba^{2+} — 1.29; Mg^{2+} — 0.65; SO_4^{2-} — 2.95; CrO_4^{2-} — 3.00.

†The term "rare elements" has not been universally accepted. According to Shcherbina, to be classed as rare, chemical elements should be present in the earth crust in amounts less than 0.01%.

Chapter 3

Law of Mass Action in Geochemistry

This law, discovered by Gouldberg in 1867, has a functional significance in the analyses of supergene processes. Let us assume that the following chemical reaction is taking place in a dilute aqueous solution:

$$nB + mC \rightleftarrows kR + lS,$$

where B, C, R, and S are participating substances (molecules or ions), and n, m, k, and l are their stoichiometric quantities.

According to the law of mass action, the rate, V_1, of direct reaction is directly proportional to the acting masses B and C. Since n molecules of B and m molecules of C are involved, the corresponding concentrations are raised to the powers n and m.

$$V_1 = K_1 [B]^n [C]^m.$$

Since the reaction is reversible, the reverse reaction should also be possible. Its rate would be:

$$V_2 = K_2 [R]^k [S]^l.$$

The state of equilibrium of this chemical solution corresponds to that instant when the rates of both direct and reverse reactions are the same ($V_2 = V_1$). Then,

$$K_1 [B]^n [C]^m = K_2 [R]^k [S]^l$$

or

$$\frac{[R]^k [S]^l}{[B]^n [C]^m} = \frac{K_1}{K_2} = K.$$

This formula expresses the law of mass action. When chemical equilibrium is attained at constant temperature and pressure, the ratio of concentration products of newly formed substances to initial substances is constant. Denoted as K, this is known as the equilibrium constant. It is independent of the concentrations of the reacting substances but varies with temperature. Its values for simpler reactions may be found in any handbook on physics and chemistry.

DISSOCIATION CONSTANT AND pH OF WATER

As is well known, water dissociates into its ions according to the equation:*

$$H_2O \rightleftharpoons H^+ + OH^-.$$

Expressing this in terms of the mass law, we get:

$$\frac{[H^+][OH]^-}{[H_2O]} = K.$$

Water dissociates only slightly at room temperature — to the amount of 10^{-7} moles/liter. Considering that one liter of water contains 55.5 moles of H_{20}, the amount of dissociation is negligible. Therefore, we can assume the concentration of undissociated H_2O molecules to be constant:

$$K[H_2O] = [H^+][OH^-].$$

In other words, the product of the concentrations of hydrogen and hydroxyl ions† is constant for the given temperature. It is known as the ionization constant of water,

$$[H^+][OH^-] = K_{H_2O}.$$

At 22°C this constant is equal to 10^{-14}. The concentration of each of these two ions is equal in neutral waters. The H^+ concentration is more than 10^{-7} in water having an acidic reaction and less than 10^{-7} in alkaline waters. For convenience, the degree of acidity–alkalinity of the given medium is expressed as the pH number which is the negative logarithm of the hydrogen ion concentration.

*The reaction of dissociation is actually:

$$H_2O + H_2O \rightleftharpoons H_3O^+ + OH-,$$

where H_3O^+ is the hydronium ion. However, it is customary to write it as the hydrogen ion, H^+.

†In physical chemistry, concentrations are given in moles/liter.

The migrational ability of chemical elements depends on the acidity or alkalinity of natural waters. Most chemical elements form their most soluble compounds in strongly acid media and their least soluble compounds in neutral media. Compounds of some elements are easily soluble in alkaline solutions with pH values of 9 to 10.

During geochemical prospecting, pH values of natural waters must be taken into account because haloes of the same chemical elements attain different dimensions and intensities under different conditions. Acid and weakly acid waters (pH under 6) favor the migration of Ca, Sr, Ba, Ra, Cu, Zn, Cd, Cr^{3+}, Fe^{2+}, Mn^{2+}, and Ni^{2+}. In alkaline solutions (at pH above 7), most of these elements move sluggishly. Hexavalent Cr, Se, and Mo and pentavalent V and As are very mobile in alkaline waters.

Values of pH for the beginning of precipitation of metallic hydroxides from dilute solutions of their salts are given in Table 7 together with the corresponding solubility products. Usually, the precipitation of hydroxides takes place at a pH higher by 0.5 to 1.5 than the pH of the beginning of precipitation. For this reason it is more proper to speak of the pH interval of precipitation, such as 10.5-11 for Mg^{2+}, 4.1-6.5 for Al^{3+}, etc.

On basis of solubility products, I. P. Onufrienok (1959) has constructed a graph relating the solubilities of metal hydroxides to pH (Fig. 11).

The relationship between the concentration in solution and precipitation is obvious. For example, Al^{3+} precipitates at pH 3.6 from a solution containing 1 g/liter; it precipitates at pH 5.3 if the solution contains only 10 mg/liter ($1 \cdot 10^{-5}$ g/liter). Such data, of course, are illustrations and should not be assumed to hold true for natural waters.

In conditions of the supergene zone, some relatively stable colloidal solutions may develop during the precipitation of metals, thus increasing the migrational ability of elements. Appreciable influences are also exerted by sorption, the formation of complex ions, etc. Nevertheless, the "precipitation values" of pH explain migrational peculiarities of elements within the supergene zone. Obviously, conditions of the precipitation of Fe^{3+} in water at pH 4 is only $2 \cdot 10^{-6}$ g/liter, i.e., very small. This cation exists in zones of oxidation of sulfidic ore deposits in very acid waters. As the pH of those waters rises to 4, iron precipitates in the form of different limonites. Al^{3+} is also present in strongly acid waters.

Table 7. pH of Metal Precipitation from Dilute Solutions and Solubility Products (at 25°C)

Hydroxide	pH	Solubility product	Metal concentration in saturated solution								
			pH = 11		pH = 10		pH = 9		pH = 8		pH = 7
			mole/liter	g/liter	mole/liter	g/liter	mole/liter	g/liter	mole/liter	g/liter	mole/liter
Sn(OH)$_4$	2	$1\cdot10^{-57}$	$1\cdot10^{-45}$	$1.2\cdot10^{-43}$	$1\cdot10^{-41}$	$1.2\cdot10^{-39}$	$1\cdot10^{-37}$	$1.2\cdot10^{-35}$	$1\cdot10^{-33}$	$1.2\cdot10^{-31}$	$1\cdot10^{-29}$
Zr(OH)$_4$	2	$8\cdot10^{-52}$	$8\cdot10^{-40}$	$7.3\cdot10^{-38}$	$8\cdot10^{-36}$	$7.3\cdot10^{-34}$	$8\cdot10^{-32}$	$7.3\cdot10^{-30}$	$8\cdot10^{-28}$	$7.3\cdot10^{-26}$	$8\cdot10^{-24}$
Th(OH)$_4$	3.5	$1\cdot10^{-50}$	$1\cdot10^{-38}$	$2.3\cdot10^{-36}$	$1\cdot10^{-34}$	$2.3\cdot10^{-32}$	$1\cdot10^{-30}$	$2.3\cdot10^{-28}$	$1\cdot10^{-26}$	$2.3\cdot10^{-24}$	$1\cdot10^{-22}$
Co(OH)$_3$	—	$2.5\cdot10^{-43}$	$2.5\cdot10^{-34}$	$1.5\cdot10^{-32}$	$2.5\cdot10^{-31}$	$1.5\cdot10^{-29}$	$2.5\cdot10^{-28}$	$1.5\cdot10^{-26}$	$2.5\cdot10^{-25}$	$1.5\cdot10^{-23}$	$2.5\cdot10^{-22}$
Sb(OH)$_3$	0.9	$4\cdot10^{-42}$	$4\cdot10^{-33}$	$4.9\cdot10^{-31}$	$4\cdot10^{-30}$	$4.9\cdot10^{-28}$	$4\cdot10^{-27}$	$4.9\cdot10^{-25}$	$4\cdot10^{-24}$	$4.9\cdot10^{-22}$	$4\cdot10^{-21}$
Fe(OH)$_3$	2.48	$4\cdot10^{-38}$	$4\cdot10^{-29}$	$2.2\cdot10^{-27}$	$4\cdot10^{-26}$	$2.2\cdot10^{-24}$	$4\cdot10^{-23}$	$2.2\cdot10^{-21}$	$4\cdot10^{-20}$	$2.2\cdot10^{-18}$	$4\cdot10^{-17}$
Ga(OH)$_3$	3.5	$5\cdot10^{-37}$	$5\cdot10^{-28}$	$3.5\cdot10^{-26}$	$5\cdot10^{-25}$	$3.5\cdot10^{-23}$	$5\cdot10^{-22}$	$3.5\cdot10^{-20}$	$5\cdot10^{-19}$	$3.5\cdot10^{-18}$	$5\cdot10^{-16}$
Al(OH)$_3$	4.1	$1.9\cdot10^{-33}$	$1.9\cdot10^{-24}$	$5.1\cdot10^{-23}$	$1.9\cdot10^{-21}$	$5.1\cdot10^{-20}$	$1.9\cdot10^{-18}$	$5.1\cdot10^{-17}$	$1.9\cdot10^{-15}$	$5.1\cdot10^{-14}$	$1.9\cdot10^{-12}$
In(OH)$_3$	3.7	$1\cdot10^{-33}$	$1\cdot10^{-24}$	$1.1\cdot10^{-22}$	$1\cdot10^{-21}$	$1.1\cdot10^{-19}$	$1\cdot10^{-18}$	$1.1\cdot10^{-16}$	$1\cdot10^{-15}$	$1.1\cdot10^{-13}$	$1\cdot10^{-12}$
Cr(OH)$_3$	5.3	$7\cdot10^{-31}$	$7\cdot10^{-22}$	$3.6\cdot10^{-20}$	$7\cdot10^{-19}$	$3.6\cdot10^{-17}$	$7\cdot10^{-16}$	$3.6\cdot10^{-14}$	$7\cdot10^{-13}$	$3.6\cdot10^{-11}$	$7\cdot10^{-10}$
Ti(OH)$_3$	1.4—1.6	$1\cdot10^{-30}$	$1\cdot10^{-21}$	$4.8\cdot10^{-20}$	$1\cdot10^{-18}$	$4.8\cdot10^{-17}$	$1\cdot10^{-15}$	$4.8\cdot10^{-14}$	$1\cdot10^{-12}$	$4.8\cdot10^{-11}$	$1\cdot10^{-9}$
Bi(OH)$_3$	4.5	$1\cdot10^{-30}$	$1\cdot10^{-21}$	$2\cdot10^{-19}$	$1\cdot10^{-18}$	$2\cdot10^{-16}$	$1\cdot10^{-15}$	$2\cdot10^{-13}$	$1\cdot10^{-12}$	$2\cdot10^{-10}$	$1\cdot10^{-9}$
Sn(OH)$_2$	3.0	$1\cdot10^{-27}$	$1\cdot10^{-21}$	$1.2\cdot10^{-19}$	$1\cdot10^{-19}$	$1.2\cdot10^{-17}$	$1\cdot10^{-17}$	$1.2\cdot10^{-15}$	$1\cdot10^{-15}$	$1.2\cdot10^{-13}$	$1\cdot10^{-13}$
Sc(OH)$_3$	4.9	$1\cdot10^{-27}$	$1\cdot10^{-18}$	$4.5\cdot10^{-17}$	$1\cdot10^{-15}$	$4.5\cdot10^{-14}$	$1\cdot10^{-12}$	$4.5\cdot10^{-11}$	$1\cdot10^{-9}$	$4.5\cdot10^{-8}$	$1\cdot10^{-6}$
Hg(OH)$_2$	7.0	$3\cdot10^{-26}$	$3\cdot10^{-20}$	$6\cdot10^{-18}$	$3\cdot10^{-18}$	$6\cdot10^{-16}$	$3\cdot10^{-16}$	$6\cdot10^{-14}$	$3\cdot10^{-14}$	$6\cdot10^{-12}$	$3\cdot10^{-12}$
Y(OH)$_3$	6.8	$1\cdot10^{-24}$	$1\cdot10^{-15}$	$8.9\cdot10^{-14}$	$1\cdot10^{-12}$	$8.9\cdot10^{-11}$	$1\cdot10^{-9}$	$8.9\cdot10^{-8}$	$1\cdot10^{-6}$	$8.9\cdot10^{-5}$	$1\cdot10^{-3}$
La(OH)$_3$	8	$1\cdot10^{-20}$	$1\cdot10^{-11}$	$1.4\cdot10^{-9}$	$1\cdot10^{-8}$	$1.4\cdot10^{-6}$	$1\cdot10^{-5}$	$1.4\cdot10^{-3}$	$1\cdot10^{-2}$	1.4	—
Be(OH)$_2$	5.7	$1\cdot10^{-20}$	$1\cdot10^{-14}$	$9\cdot10^{-14}$	$1\cdot10^{-12}$	$9\cdot10^{-12}$	$1\cdot10^{-10}$	$9\cdot10^{-10}$	$1\cdot10^{-8}$	$9\cdot10^{-8}$	$1\cdot10^{-6}$
Ni(OH)$_2$	6.7	$8.7\cdot10^{-19}$	$8.7\cdot10^{-13}$	$5\cdot10^{-11}$	$8.7\cdot10^{-11}$	$5\cdot10^{-9}$	$8.7\cdot10^{-9}$	$5\cdot10^{-7}$	$8.7\cdot10^{-7}$	$5\cdot10^{-5}$	$8.7\cdot10^{-5}$
Cu(OH)$_2$	5.4	$1.6\cdot10^{-19}$	$1.6\cdot10^{-13}$	$1\cdot10^{-11}$	$1.6\cdot10^{-11}$	$1\cdot10^{-9}$	$1.6\cdot10^{-9}$	$1\cdot10^{-7}$	$1.6\cdot10^{-7}$	$1\cdot10^{-5}$	$1.6\cdot10^{-5}$
Zn(OH)$_2$	5.2	$4.5\cdot10^{-17}$	$4.5\cdot10^{-11}$	$2.9\cdot10^{-9}$	$4.5\cdot10^{-9}$	$2.9\cdot10^{-7}$	$4.5\cdot10^{-7}$	$2.9\cdot10^{-5}$	$4.5\cdot10^{-5}$	$2.9\cdot10^{-3}$	$4.5\cdot10^{-3}$
Fe(OH)$_2$	5.5	$4.8\cdot10^{-16}$	$4.8\cdot10^{-10}$	$2.7\cdot10^{-8}$	$4.8\cdot10^{-8}$	$2.7\cdot10^{-6}$	$4.8\cdot10^{-6}$	$2.7\cdot10^{-4}$	$4.8\cdot10^{-4}$	$2.7\cdot10^{-2}$	$4.8\cdot10^{-2}$
Pb(OH)$_2$	6.0	$7\cdot10^{-16}$	$7\cdot10^{-10}$	$1.4\cdot10^{-7}$	$7\cdot10^{-8}$	$1.4\cdot10^{-5}$	$7\cdot10^{-6}$	$1.4\cdot10^{-3}$	$7\cdot10^{-4}$	$1.4\cdot10^{-1}$	$7\cdot10^{-2}$
Co(OH)$_2$	6.8	$1.3\cdot10^{-15}$	$1.3\cdot10^{-9}$	$7.6\cdot10^{-8}$	$1.3\cdot10^{-7}$	$7.6\cdot10^{-6}$	$1.3\cdot10^{-5}$	$7.6\cdot10^{-4}$	$1.3\cdot10^{-3}$	$7.6\cdot10^{-2}$	$1.3\cdot10^{-1}$
Mn(OH)$_2$	9.0	$4.0\cdot10^{-14}$	$4\cdot10^{-8}$	$2.2\cdot10^{-6}$	$4\cdot10^{-6}$	$2.2\cdot10^{-4}$	$4\cdot10^{-4}$	$2.2\cdot10^{-2}$	$4\cdot10^{-2}$	2.2	4
Cd(OH)$_2$	6.7	$2.3\cdot10^{-14}$	$2.3\cdot10^{-8}$	$2.6\cdot10^{-6}$	$2.3\cdot10^{-6}$	$2.6\cdot10^{-4}$	$2.3\cdot10^{-4}$	$2.6\cdot10^{-2}$	$2.3\cdot10^{-2}$	2.6	—
Mg(OH)$_2$	10.5	$5\cdot10^{-12}$	$5\cdot10^{-6}$	$1.2\cdot10^{-4}$	$5\cdot10^{-4}$	$1.2\cdot10^{-2}$	$5\cdot10^{-2}$	1	—	—	—
AgOH	9.0	$2\cdot10^{-8}$	$2\cdot10^{-5}$	$2.2\cdot10^{-3}$	$2\cdot10^{-4}$	$2.2\cdot10^{-2}$	$2\cdot10^{-3}$	$2.2\cdot10^{-1}$	$2\cdot10^{-2}$	2.1	$2\cdot10^{-1}$
Ce(OH)$_3$	7.4	—	—	—	—	—	—	—	—	—	—
Nd(OH)$_3$	7	—	—	—	—	—	—	—	—	—	—
UO$_2$(OH)$_2$	4.2	—	—	—	—	—	—	—	—	—	—
NbO$_2$OH	0.4	—	—	—	—	—	—	—	—	—	—

Table 7 (continued)

Hydroxide	Metal concentration in saturated solution										
	pH = 7	pH = 6		pH = 5		pH = 4		pH = 3		pH = 2	
	g/liter	mole/liter	g/liter	mole/liter	g/liter	mole/liter	g/liter	mole/liter	g/liter	mole/liter	g/liter
Sn(OH)$_4$	$1.2 \cdot 10^{-27}$	$1 \cdot 10^{-25}$	$1.2 \cdot 10^{-23}$	$1 \cdot 10^{-21}$	$1.2 \cdot 10^{-19}$	$1 \cdot 10^{-17}$	$1.2 \cdot 10^{-15}$	$1 \cdot 10^{-13}$	$1.2 \cdot 10^{-11}$	$1 \cdot 10^{-9}$	$1.2 \cdot 10^{-7}$
Zr(OH)$_4$	$7.3 \cdot 10^{-22}$	$8 \cdot 10^{-20}$	$7.3 \cdot 10^{-18}$	$8 \cdot 10^{-16}$	$7.3 \cdot 10^{-14}$	$8 \cdot 10^{-12}$	$7.3 \cdot 10^{-10}$	$8 \cdot 10^{-8}$	$7.3 \cdot 10^{-6}$	$8 \cdot 10^{-4}$	$7.3 \cdot 10^{-2}$
Th(OH)$_4$	$2.3 \cdot 10^{-20}$	$1 \cdot 10^{-18}$	$2.3 \cdot 10^{-16}$	$1 \cdot 10^{-14}$	$2.3 \cdot 10^{-12}$	$1 \cdot 10^{-10}$	$2.3 \cdot 10^{-8}$	$1 \cdot 10^{-6}$	$2.3 \cdot 10^{-4}$	$1 \cdot 10^{-2}$	2.3
Co(OH)$_3$	$1.5 \cdot 10^{-20}$	$2.5 \cdot 10^{-19}$	$1.5 \cdot 10^{-17}$	$2.5 \cdot 10^{-16}$	$1.5 \cdot 10^{-14}$	$2.5 \cdot 10^{-13}$	$1.5 \cdot 10^{-11}$	$2.5 \cdot 10^{-10}$	$1.5 \cdot 10^{-8}$	$2.5 \cdot 10^{-7}$	$1.5 \cdot 10^{-5}$
Sb(OH)$_3$	$4.9 \cdot 10^{-19}$	$4 \cdot 10^{-18}$	$4.9 \cdot 10^{-16}$	$4 \cdot 10^{-15}$	$4.9 \cdot 10^{-13}$	$4 \cdot 10^{-12}$	$4.9 \cdot 10^{-10}$	$4 \cdot 10^{-9}$	$4.9 \cdot 10^{-7}$	$4 \cdot 10^{-6}$	$4.9 \cdot 10^{-4}$
Fe(OH)$_3$	$2.2 \cdot 10^{-15}$	$4 \cdot 10^{-14}$	$2.2 \cdot 10^{-12}$	$4 \cdot 10^{-11}$	$2.2 \cdot 10^{-9}$	$4 \cdot 10^{-8}$	$2.2 \cdot 10^{-6}$	$4 \cdot 10^{-5}$	$2.2 \cdot 10^{-3}$	$4 \cdot 10^{-2}$	2.2
Ga(OH)$_3$	$3.5 \cdot 10^{-14}$	$5 \cdot 10^{-13}$	$3.5 \cdot 10^{-11}$	$5 \cdot 10^{-10}$	$3.5 \cdot 10^{-8}$	$5 \cdot 10^{-7}$	$3.5 \cdot 10^{-5}$	$5 \cdot 10^{-4}$	$3.5 \cdot 10^{-2}$	$5 \cdot 10^{-1}$	35
Al(OH)$_3$	$5.1 \cdot 10^{-11}$	$1.9 \cdot 10^{-9}$	$5.1 \cdot 10^{-8}$	$1.9 \cdot 10^{-6}$	$5.1 \cdot 10^{-5}$	$1.9 \cdot 10^{-3}$	$5 \cdot 10^{-2}$	—	—	—	—
In(OH)$_3$	$1.1 \cdot 10^{-10}$	$1 \cdot 10^{-9}$	$1.1 \cdot 10^{-7}$	$1 \cdot 10^{-6}$	$1.1 \cdot 10^{-4}$	$1 \cdot 10^{-3}$	$1.1 \cdot 10^{-1}$	—	—	—	—
Cr(OH)$_3$	$3.6 \cdot 10^{-8}$	$7 \cdot 10^{-7}$	$3.6 \cdot 10^{-5}$	$7 \cdot 10^{-4}$	$3.6 \cdot 10^{-2}$	$7 \cdot 10^{-1}$	3.6	—	—	—	—
Ti(OH)$_4$	$4.8 \cdot 10^{-1}$	—	—	—	—	—	—	—	—	—	—
Bi(OH)$_3$	$2 \cdot 10^{-7}$	$1 \cdot 10^{-6}$	$2 \cdot 10^{-4}$	$1 \cdot 10^{-3}$	$2 \cdot 10^{-1}$	—	—	—	—	—	—
Sn(OH)$_2$	$1.2 \cdot 10^{-11}$	$1 \cdot 10^{-11}$	$1.2 \cdot 10^{-9}$	$1 \cdot 10^{-9}$	$1.2 \cdot 10^{-7}$	$1 \cdot 10^{-7}$	$1.2 \cdot 10^{-5}$	$1 \cdot 10^{-5}$	$1.2 \cdot 10^{-3}$	$1 \cdot 10^{-3}$	$1.2 \cdot 10^{-1}$
Sc(OH)$_3$	$4.5 \cdot 10^{-5}$	$1 \cdot 10^{-3}$	$4.5 \cdot 10^{-2}$	1	0.1	—	—	—	—	—	—
Hg(OH)$_2$	$6 \cdot 10^{-10}$	$3 \cdot 10^{-10}$	$6 \cdot 10^{-8}$	$3 \cdot 10^{-8}$	$6 \cdot 10^{-6}$	$3 \cdot 10^{-6}$	$6 \cdot 10^{-4}$	$3 \cdot 10^{-4}$	$6 \cdot 10^{-2}$	$3 \cdot 10^{-2}$	6
Y(OH)$_3$	$8.9 \cdot 10^{-2}$	1	89	—	—	—	—	—	—	—	—
La(OH)$_3$	—	—	—	—	—	—	—	—	—	—	—
Be(OH)$_2$	$9 \cdot 10^{-4}$	$1 \cdot 10^{-4}$	$9 \cdot 10^{-2}$	$1 \cdot 10^{-2}$	9	—	—	—	—	—	—
Ni(OH)$_2$	$5 \cdot 10^{-3}$	$8.7 \cdot 10^{-3}$	$5 \cdot 10^{-1}$	$8.7 \cdot 10^{-1}$	50	—	—	—	—	—	—
Cu(OH)$_2$	$1 \cdot 10^{-3}$	$1.6 \cdot 10^{-3}$	$1 \cdot 10^{-1}$	$1.6 \cdot 10^{-1}$	10	—	—	—	—	—	—
Zn(OH)$_2$	$2.9 \cdot 10^{-1}$	$4.5 \cdot 10^{-1}$	29	—	—	—	—	—	—	—	—
Fe(OH)$_2$	2.7	4.8	—	—	—	—	—	—	—	—	—
Pb(OH)$_2$	—	—	—	—	—	—	—	—	—	—	—
Co(OH)$_2$	7.6	—	—	—	—	—	—	—	—	—	—
Mn(OH)$_2$	—	—	—	—	—	—	—	—	—	—	—
Cd(OH)$_2$	—	—	—	—	—	—	—	—	—	—	—
Mg(OH)$_2$	—	—	—	—	—	—	—	—	—	—	—
AgOH	—	—	—	—	—	—	—	—	—	—	—
Ce(OH)$_3$	—	—	—	—	—	—	—	—	—	—	—
Nd(OH)$_3$	—	—	—	—	—	—	—	—	—	—	—
UO$_2$(OH)$_2$	—	—	—	—	—	—	—	—	—	—	—
NbO$_2$OH	—	—	—	—	—	—	—	—	—	—	—

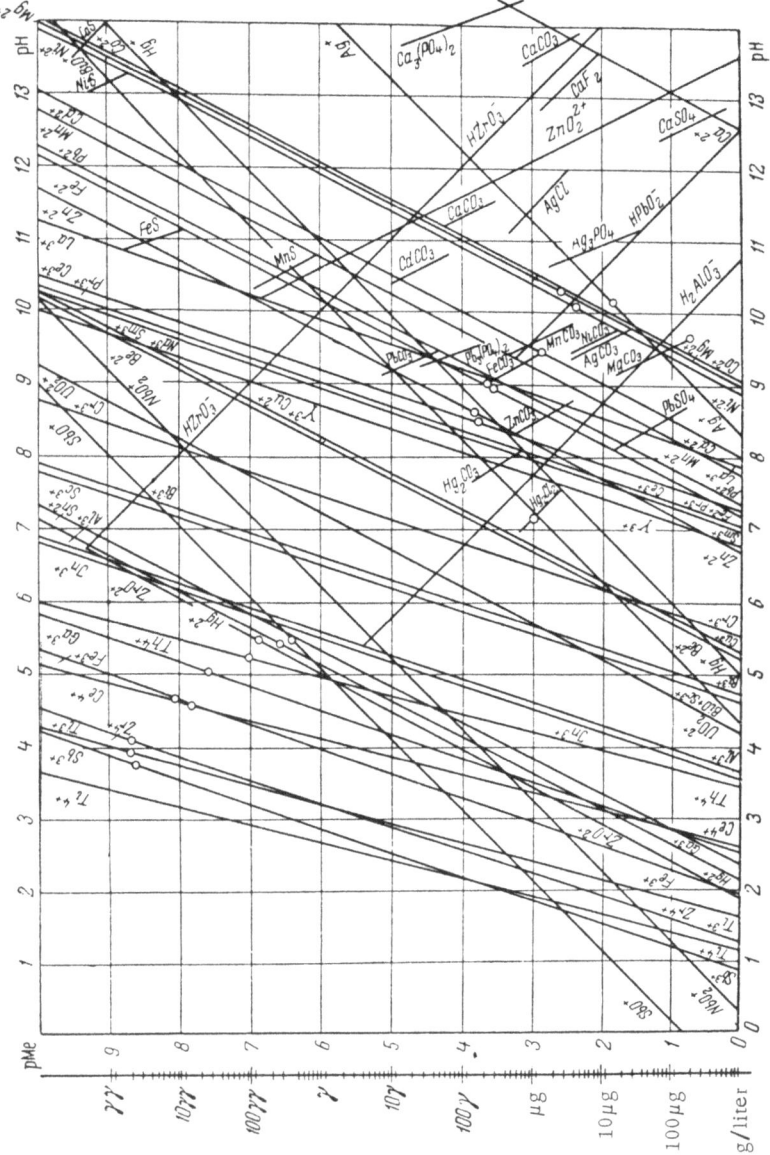

Fig. 11. Solubility limits for metal hydroxides in relation to the magnitude of the pH (after Onufrienok, 1959).

At pH 5, its concentration is lowered to $5 \cdot 10^{-5}$ g/liter, which means a negligible solubility in comparison with its high abundance. Trivalent Co, Cr, Bi, Sb, and Sc, bivalent Sn, and tetravalent Th, Zr, and Ti can exist only in very acid waters which are uncommon in the earth's crust. Therefore, these cations precipitate out of natural waters and possess low migrational ability. On the other hand, monovalent Ag, bivalent Ni, Co, Zn, Cd, and Pb and trivalent V and La remain dissolved in significant amounts even at pH 8 concentrations being 2.2 g/liter for Cd^{2+}, 2.1 for Ag^+, $8.7 \cdot 10^{-1}$ for Pb^{2+}, $1.5 \cdot 10^{-2}$ for Co^{2+}, $9.3 \cdot 10^{-3}$ for Ni^{2+}, $2.9 \cdot 10^{-3}$ for Zn^{2+}, etc. Their contents in natural waters are still lower and hydroxides do not precipitate. Therefore, pH should not directly affect their migration. Their precipitation is achieved through the formation of insoluble compounds (phosphates, arsenates, carbonates, etc.) or through sorption. However, the effect of pH might be indirect. In mineralized regions where metals locally reach large concentrations, the precipitation of certain hydroxides might be feasible.

Elements like Na, K, Ca, Rb, Cs, and Sr do not form hydroxides under the conditions existing in the earth's crust, and the pH of water can affect their precipitation only indirectly.

Thus evaluating the role of pH of water as a factor in migration the approach should be made selectively on the basis of hydroxide solubilities, elements' abundances, and their content in water. The role of pH is negligible in the precipitation of the hydroxides of many rare elements because of their extremely low concentration.

DISSOCIATION CONSTANT OF ACIDS

Data on the dissociation of acids encountered in natural waters are very important to geochemists.

Let us consider the constant of dissociation of carbonic acid. Like any other two-base acid, it may dissociate in two ways:

$$H_2CO_3 \rightleftharpoons H^+ + HCO_3^-,$$
$$HCO_3^- \rightleftharpoons H^+ + CO_3^{2-}.$$

In accordance with the law of mass action we can write:

$$\frac{[H^+][HCO_3^-]}{[H_2CO_3]} = K_1$$

and

$$\frac{\left[CO_3^{2-}\right][H^+]}{\left[HCO_3^-\right]} = K_2.$$

At 25° C, $K_1 = 4 \cdot 10^{-7}$ and $K_2 = 5.6 \cdot 10^{-11}$. Such low values indicate carbonic acid to be a weak acid (compare with $K_2 = 1.2 \cdot 10^{-2}$ for sulfuric acid). Since the concentration of the hydrogen ion enters each equation, the pH of water would change depending on the concentration of H_2CO_3, HCO_3^-, and CO_3^{2-}. Relationships between different forms of carbonic acid and the corresponding pH may be calculated in a fairly simple way (Table 8). Considering that bicarbonate waters predominate within the supergene zone of the continents, considerable geochemical significance is attached to such relationships. For surface and ground waters of steppes and deserts where pH fluctuates about a value of 8, the bicarbonate ion, HCO_3^-, is predominant. The CO_3^{2-} ion is not even common there, since strongly alkaline waters are rare in those supergene zones. Undissociated H_2CO_3 predominates in weakly acid (pH 5-6) waters in humid regions. Because of very low pH, both HCO_3^- and CO_3^{2-} ions are often absent in oxidation zones of sulfidic ore deposits.

Hydrogen sulfide is another weak acid important in geochemistry. At 20° C, we have for it $K_1 = 8.7 \cdot 10^{-8}$ and $K_2 = 3.6 \cdot 10^{-13}$. From these data it is evident that hydrogen sulfide is an even weaker acid than carbonic. In fact, its dissociation is conditioned by the carbonatic equilibrium of natural water. Relationships between pH and the dissociation products of hydrogen sulfide are given in Table 9. It will be noted that the S^{2-} ion should be practically absent from the supergene zone. H_2S should predominate in neutral and acid waters and HS^- in alkaline waters.

Table 8. Relative Amounts (% molar) of the Dissociation Products of Carbonic Acid Dissolved in Water of Various ph Values (Alekin, 1953)

H_2CO_3 derivative	pH							
	4	5	6	7	8	9	10	11
$[H_2CO_3]$	99.7	97.0	76.7	24.29	3.22	0.32	0.02	—
$\left[HCO_3^-\right]$	0.3	3.0	23,3	74.98	96.70	95.84	71.43	20.0
$\left[CO_3^{2-}\right]$	—	—	—	0.03	0.08	3.84	28.55	80.0

Table 9. Relative Amounts (% molar) of the Dissociation Products of H_2S Dissolved in Water of Various ph Values (Alekin, 1953)

H_2S derivative	pH						
	4	5	6	7	8	9	10
$[H_2S]$	99.91	99.1	91.66	52.35	9.81	1.09	0.11
$[HS^-]$	0.09	0.90	8.34	47.65	90.19	98.91	99.89
$[S^{2-}]$	—	—	—	—	—	—	0.002

Siliceous acid is generally present in natural waters in undissociated form, the $HSiO_3^-$ ion appearing only in alkaline media (Table 10).

Orthophosphoric acid is an acid of intermediate strength. At 20° C, its dissociation constants are $K_1 = 7.9 \cdot 10^{-3}$, $K_2 = 1 \cdot 10^{-7}$, and $K_3 = 4.5 \cdot 10^{-12}$. Its dissociation constants are given in Table 11 for different pH values. It follows that undissociated H_3PO_4 may be encountered in appreciable amounts only in strongly acid environments of oxidizing sulfides, where $(PO_4)^{3-}$ would be practically absent. The $(H_2PO_4)^-$ ion predominates in weakly acid waters and $(HPO_4)^{2-}$ in neutral and weakly alkaline waters.

Strong acids are not characteristic of the supergene zone and sulfuric acid is found only in the oxidation zones of sulfide ore deposits in oxygenated ground waters which drain from pyritic sedimentary rocks. The pH of such waters drops low, to 2 and even to 1. Another strong acid—hydrochloric— is rarely found except in certain volcanic regions (K. K. Zelenov).

Arsenic, vanadium, boron, selenium, and other elements can also form acids under appropriate conditions. However, because of the low abundances of these elements their concentrations are generally low. Their dissociation depends entirely on the pH produced by dissociation of the principal components—carbonic

Table 10. Relative Amounts (% molar) of the Dissociation Products of Siliceous Acid Dissolved in Water of Various ph Values (Alekin, 1953)

H_2SiO_3 derivative	pH				
	7	8	9	10	11
$[H_2SiO_3]$	99.6	96.1	71.5	20.0	2.4
$[HSiO_3^-]$	0.4	3.9	28.5	80.0	97.6

Table 11. Relative Amounts (% molar) of the Dissociation
Products of Phosphoric Acid Dissolved in Water of Various
ph Values (Alekin, 1953)

H_3PO_4 derivative	pH							
	5	6	7	8	8,5	9	10	11
$[H_3PO_4]$	0.1	0.01	—	—	—	—	—	—
$[H_2PO_4^-]$	97.99	83.67	33.90	4.88	1.60	0.51	0.05	—
$[HPO_4^{2-}]$	1.91	16.32	66.10	95.12	98.38	99.45	99.59	96.53
$[PO_4^{3-}]$	—	—	—	—	0.01	0.04	0.36	3.47

acid, carbonates, sulfuric acid, etc. Orthoarsenic acid, H_3AsO_4, is formed in the oxidation zone of ore deposits rich in arsenides. Its strength is similar to that of orthophosphoric acid: $K_1 = 5.6 \cdot 10^{-3}$, $K_2 = 1.7 \cdot 10^{-7}$, $K_3 = 3 \cdot 10^{-12}$. Boric acid, H_3BO_3, is weaker than carbonic, its constant being (at 25° C): $K_1 = 5.8 \cdot 10^{-10}$.

Hydrogen selenide, H_2Se, dissociates in a manner similar to hydrogen sulfide and produces ions HSe^- and Se^{2-}. Having $K_1 = n \cdot 10^{-4}$ and $K_2 = n \cdot 10^{-11}$, it is a stronger acid than hydrogen sulfide.

Humic, fulvic, and other organic acids may develop in certain environments of the supergene zone. As a rule, these are weak, polybase acids with poorly known dissociation constants.

COMPLEX IONS

Most metals present in natural waters occur there in the form of hydroxyl complexes, polymer ions, and complex anions (Brusilovskii, 1963). Hexavalent uranium is a good example in this respect.

The complex uranyl cation $(UO_2)^{2+}$ seldom produces hydroxyl complexes or complex anions. Therefore, depending on the general aspect of mineralization, on the pH, and on the uranium concentration, the following ions may be present in natural waters: UO_2^{2+}, $UO_2(OH)^+$, $[UO_2(CO_3)_2(H_2O)_2]^{2-}$, $[UO_2(CO_3)_3]^{4-}$, and undissociated molecules $UO_2(OH)_2$.

Copper, silver, yttrium, and yttrian rare earths form carbonate complexes which are similar to those of uranium.

For example, basic carbonates of copper are difficultly soluble in neutral media; this explains the existence of such minerals as malachite, $CuCO_3 \cdot Cu(OH)_2$, and azurite, $2CuCO_3 \cdot Cu(OH)_2$. These salts dissolve in solutions of alkali carbonates and bicarbonates and produce complexes of the type $Na_2Cu(CO_3)_2$, $NaHCu(CO_3)_2$, etc., with copper being a part of the complex anion.

Shcherbina and Ignatova (1955) have demonstrated that when a precipitate of copper carbonate is treated with a concentrated solution of sodium bicarbonate (which should contain 8 to 9 g/liter $NaHCO_3$), copper is dissolved, coloring the solution blue. Copper phosphates (elite) also dissolve in concentrated solutions of sodium carbonate and the CuO content of such a solution reaches 0.166 g per liter and that of P_2O_5 0.118 g/liter.

Blue solutions of copper complexes diffuse through semipermeable membranes (parchment), proving that the solution is true and not colloidal.

The so-called chelation complexes comprise a very peculiar group of compounds in which a molecule of organic matter "grabs," as it were, an inorganic ion. Such complexes are formed by iron, zinc, manganese, copper, and other metals. Some chelates are poorly soluble in water because the formation of such inner complexes lowers the migrational ability of metals. There are, however, many chelates which are soluble in water.*

Chelates decompose in light with the separation of hydroxides of heavy metals. They play a large role in the lives of plants and animals, participating in many physiological processes. This group includes chlorophyll, hemoglobin, enzymes, many fermenting agents, metal humates, etc. Since chelates extract metals out of minerals and render these mobile, many authors hold that weathering is related to chelation.

In the opinion of Antipov-Karataev and Tsyuryupa (1961), iron and aluminum migrate through podzol soils in chelate forms. This is confirmed by the experiments of Kaurichev, Kulakov, and Nozdrukova (1958) which showed that soluble organic compounds of bivalent iron form in swamp soils.

Metals of chelates can be exchanged for those of soils. Thus, these metals are available to plants.

---

*As an example of such a complex we cite the copper salt of amino-acetic acid:

$$CO\!-\!O \diagdown \quad \diagup O \!-\!\!-\!\!-\! CO$$
$$| \qquad \diagup Cu \diagdown \qquad |$$
$$CH_2\!-\!NH_2 \quad NH_2\!-\!CH_2$$

Complexes of the type $(NaSO_4)^-$ exist in sea water, and for that matter in any strongly mineralized water. Brusilovskii estimated that this complex is second only to the Cl^- ion, while free SO_4^{2-} ions are rare. Chemical analyses of natural waters are usually expressed in terms of simple ions, such as Na^+, K^+, Ca^{2+}, Cl^-, SO_4^{2-}, and HCO_3^-, which do not portray the actual composition of water. Attention was first called to this fact by Vernadskii in his "History of Natural Waters" and was recently reiterated by Brusilovskii (1963).

The formation of complex ions drastically changes the conditions of precipitation of many metals. It requires a very delicate adjustment of such characteristics as the pH of hydroxide precipitation. For example, the pH of precipitation of uranyl hydroxide, $UO_2(OH)_2$, varies between 3.8 and 6.0 depending on the concentration of uranium in solution. It would seem that uranium should not migrate in neutral and alkaline waters, i.e., at pH above 6. Nevertheless, uranium does migrate in those waters, having formed the already mentioned soluble carbonate complexes $[UO_2(CO_3)_3]^{4-}$ and $[UO_2(CO_3)_2(H_2O)_2]^{2-}$. For most metals the formation of complex ions raises the pH of hydroxide precipitation and its solubility.

Complex compounds dissociate in two stages: (1) dissociation into complex and simple ions, and (2) dissociation of the complex ion itself. For example, the dissociation of the $Na_4[(UO_2)(CO_3)_3]$ complex proceeds as follows:

$$Na_4[UO_2(CO_3)_3] \rightleftarrows 4Na^+ + [UO_2(CO_3)_3]^{4-},$$

$$[UO_2(CO_3)_3]^{4-} \rightleftarrows UO_2^{2+} + 3CO_3^{2-}.$$

Rewriting the last equation in a form expressing the law of mass action* we get

$$\frac{\left[(UO_2)(CO_3)_3\right]^{4-}}{\left[UO_2^{2+}\right]\left[CO_3^{2-}\right]^3} = K = 10^{18.9}.$$

The constant K characterizes the degree of stability of the given complex ion and is known as the constant of stability. Its reciprocal, the constant of instability, is used more frequently in the chemistry of complex compounds. The greater the constant of stability or the

*L. MacClein, E. Ballwinkle, and J. Huygens, in Chemistry of Nuclear Fuels, IL, Moscow (1956).

smaller the constant of instability, the more stable is the given ion and the weaker would be its further dissociation.

Using equilibria constants available in the literature for different uranium compounds, Lisitsin (1962) calculated the forms of uranium occurring in the ground waters investigated by him. At 20° C and pH 6.6, these waters contained the following:

$$U \ldots \ldots \ldots .4 \cdot 10^{-6} \text{ g/liter } (1.7 \cdot 10^{-8} \text{ gram-atom/liter})$$
$$Cl^- \ldots \ldots .1.028 \text{ g/liter}$$
$$SO_4^{2-} \ldots \ldots .0.106 \text{ g/liter}$$
$$HCO_3^- \ldots \ldots .1.391 \text{ g/liter}$$
$$(Na^+, K^+) \ldots .0.749 \text{ g/liter}$$
$$Ca^{2+} \ldots \ldots .0.311 \text{ g/liter}$$
$$Mg^{2+} \ldots \ldots .0.071 \text{ g/liter}$$

Calculations also showed that in the given water there are practically no compounds of U^{4+}, while carbonate complexes of U^{6+} exist in the relative amounts:

$$[UO_2 (CO_3)_2 (H_2O_2)]^{2-} = 82\%$$
$$[UO_2 (CO_3)_3]^{4-} = 18\%.$$

UNDISSOCIATED MOLECULES

Undissociated molecules are common in natural waters. According to Brusilovskii in a saturated solution of gypsum at 5° C there are $2\frac{1}{2}$ times as many undissociated $CaSO_4$ molecules as Ca^{2+} ions.

The reaction of dissociation of ferric hydroxide is generally written as:

$$Fe(OH)_3 \rightleftharpoons Fe^{3+} + 3OH^-.$$

It can be calculated, using the constant of dissociation, that the concentration of Fe^{3+} ions in neutral water is only 10^{-16} to 10^{-18} mole/liter. At the same time, truly dissolved undissociated molecules of $Fe(OH)_3$ predominate. According to the same author, neutral water should contain for each Fe^{3+} ion: $3 \cdot 10^5$ ions $Fe(OH)^{2+}$, $6 \cdot 10^6$ ions $(Fe(OH)_2)^+$, and $9 \cdot 10^6$ undissociated $Fe(OH)_3$. Truly dissolved, undissociated metal hydroxides are probably the most important mode of migration of many elements (Brusilovskii, 1963).

SOLUBILITY PRODUCT

A solution of some difficultly soluble salt in equilibrium with an excess of that salt is said to be a saturated solution. As an example let us consider the reaction of precipitation of calcium fluoride:

$$Ca^{2+} + 2F^- \rightleftharpoons \underline{CaF_2} \text{ solid.}$$
$$\downarrow$$

In accordance with the law of mass action, we write:

$$\frac{[Ca^{2+}][F^-]^2}{[CaF_2] \text{ solid}} = K_1$$

or

$$[Ca^{2+}][F^-]^2 = K_1 [CaF_2] \text{ solid.}$$

Since the concentration of a substance in its solid phase is constant, we can simplify this to:

$$[Ca^{2+}][F^-]^2 = K.$$

In other words, the product of molar concentrations of ions of a given mineral in its saturated solutions is a constant called the solubility product (SP).*

The solubility product is constant for a given temperature and pressure. Let us consider its use in a concrete example, the solubility of CaF_2. Since the solubility equals $10^{-9.8}$,† the amount of Ca^{2+} and F^- ions in a saturated solution may vary but the product of the molar concentration of Ca^{2+} and the square of the molar concentration of F^- will remain the same $(10^{-9.8})$ for 25° C. Let the concentration of Ca^{2+} in water be 10^{-2} mole/liter (0.40 g/liter). Then, the concentration of F^- in that solution will be:

$$[F^-] = \sqrt{\frac{10^{-9.8}}{10^{-2}}} = 10^{-3.9}.$$

When the concentration of Ca^{2+} decreases by one-fourth the concentration of F^- will double, etc.

It follows from the rule of solubility products that when a more

---

*If several identical ions are formed from the molecules of a substance in solution, it follows from the law of mass action that the number of such ions will enter into the calculation of the SP as an exponent. Thus, the SP for $CaCO_3 = [Ca^{2+}] [CO_3^{2-}]$, for $CaF_2 = [Ca^{2+}] [F^-]^2$, and for $Fe(OH)_3 = [Fe^{+3}] [OH^-]^3$.

†See p. 74 for the calculation of the SP.

soluble salt is added to a solution of some salt, the two having some ions in common, the solubility of the first salt is lowered. For example, the solubility of CaF_2 would be lowered on the addition of $CaSO_4$. The solubility of anglesite, $PbSO_4$, would be lowered on the addition of, say, $ZnSO_4$. It is as if the added salt keeps anglesite from going into solution. Upon leaching of the added sulfate, the solubility of anglesite returns to normal.

The solubility products of various compounds are given below.

Solubility Products of Some Compounds at 25°C

Sulfides

$Bi_2S_3 - 1.6 \cdot 10^{-72}$	$Sb_2S_5 - 1 \cdot 10^{-30}$
$HgS - 4 \cdot 10^{-58}$	$NiS^* - 1.4 \cdot 10^{-24}$
$CuS - 8 \cdot 10^{-37}$	$CdS - 1 \cdot 10^{-29}$
$Cu_2S - 2.5 \cdot 10^{-50}$	$CoS^* - 2 \cdot 10^{-27}$
$Ag_2S - 1 \cdot 10^{-51}$	$FeS - 4 \cdot 10^{-19}$
$PbS - 1 \cdot 10^{-29}$	$MnS^* - 1.4 \cdot 10^{-15}$
$Sb_2S_3 - 4 \cdot 10^{-29}$	(compacted)
	$ZnS^* - 8 \cdot 10^{-26}$

Sulfates

$BaSO_4 - 1.1 \cdot 10^{-10}$	$SrSO_4 - 2.8 \cdot 10^{-7}$
$PbSO_4 - 2 \cdot 10^{-8}$	$Ag_2SO_4 - 7.7 \cdot 10^{-5}$
$Hg_2SO_4 - 5 \cdot 10^{-7}$	$CaSO_4 - 6.1 \cdot 10^{-5}$

Carbonates

$Hg_2CO_3 - 9 \cdot 10^{-17}$	$CuCO_3 - 1.4 \cdot 10^{-10}$
$PbCO_3 - 1.5 \cdot 10^{-13}$	$SrCO_3 - 1 \cdot 10^{-9}$
$CdCO_3 - 2.5 \cdot 10^{-14}$	$CaCO_3 - 4.8 \cdot 10^{-9}$
$Ag_2CO_3 - 6.2 \cdot 10^{-12}$	$BaCO_3 - 8.0 \cdot 10^{-9}$
$CoCO_3 - 1 \cdot 10^{-12}$	$NiCO_3 - 1.4 \cdot 10^{-7}$
$ZnCO_3 - 6 \cdot 10^{-11}$	$MgCO_3 - 1 \cdot 10^{-5}$
$FeCO_3^* - 2.5 \cdot 10^{-11}$	$Li_2CO_3 - 1.7 \cdot 10^{-3}$
$MnCO_3 - 1 \cdot 10^{-10}$	

Phosphates, Arsenates, and Vanadates

$Pb_3(PO_4)_2 - 8.2 \cdot 10^{-43}$	$PbHPO_4 - 1 \cdot 10^{-11}$
$Zn_3(PO_4)_2 - 9.1 \cdot 10^{-33}$	$FeHPO_4 - 4 \cdot 10^{-10}$
$Ca_3(PO_4)_2 - 3.5 \cdot 10^{-33}$	$CaHPO_4 - 5 \cdot 10^{-6}$
$Ba_3(PO_4)_2 - 1.3 \cdot 10^{-29}$	$AlPO_4 - 1 \cdot 10^{-6}$
$Ag_3PO_4 - 1.3 \cdot 10^{-20}$	$Ag_3AsO_4 - 1 \cdot 10^{-22}$
$FePO_4 - 1.3 \cdot 10^{-22}$	

The solubility products rule can only be quantitatively applied to minerals with solubilities of less than 0.01 mole/liter (roughly,

Table 12. Solubility of Metallic Compounds (after Krauskopf, 1959, with additions)

Cation	CO_3^{2-}		S^{2-}		SO_4^{2-}		Cl^-		PO_4^{3-}	
	10^{-4} mole/liter	$6\cdot10^{-3}$ g/liter	10^{-10} mole/liter	$3.2\cdot10^{-9}$ g/liter	10^{-2} mole/liter	$9.6\cdot10^{-1}$ g/liter	10^{-1} mole/liter	3.5 g/liter	10^{-9} mole/liter	$9.5\cdot10^{-7}$ g/liter
Ag^+	$3\cdot10^{-4}$	$3.21\cdot10^{-2}$	$7\cdot10^{-21}$	$7.5\cdot10^{-19}$	$3\cdot10^{-2}$	3.21	$2.8\cdot10^{-9}$	$2.9\cdot10^{-7}$	—	—
Cu^+	—	—	$3\cdot10^{-20}$	$1.9\cdot10^{-18}$	—	—	$1.8\cdot10^{-7}$	$1.8\cdot10^{-5}$	—	—
Tl^+	—	—	$1.1\cdot10^{-7}$	$2.2\cdot10^{-5}$	—	—	$1.9\cdot10^{-3}$	$3.9\cdot10^{-1}$	—	—
BiO^+	—	—	$1.3\cdot10^{-21}$	$2.9\cdot10^{-19}$	—	—	—	—	—	—
Cu^{2+}	$2.5\cdot10^{-6}$	$1.6\cdot10^{-4}$	$8\cdot10^{-27}$	$5\cdot10^{-25}$	—	—	—	—	—	—
Be^{2+}	—	—	—	—	—	—	—	—	—	—
Ca^{2+}	$4.8\cdot10^{-5}$	$1.9\cdot10^{-3}$	—	—	—	—	—	—	$2\cdot10^{-5}$	$8\cdot10^{-4}$
Zn^{2+}	$6.0\cdot10^{-7}$	$3.9\cdot10^{-5}$	$4.5\cdot10^{-14}$	$2.9\cdot10^{-12}$	—	—	—	—	—	—
Sr^{2+}	$7\cdot10^{-6}$	$6\cdot10^{-4}$	—	—	$7.6\cdot10^{-5}$	$6.6\cdot10^{-3}$	—	—	$5\cdot10^{-5}$	$4.3\cdot10^{-3}$
Cd^{2+}	$5.2\cdot10^{-8}$	$5.8\cdot10^{-6}$	$1.0\cdot10^{-18}$	$1.1\cdot10^{-16}$	—	—	—	—	$2\cdot10^{-7}$	$2.7\cdot10^{-5}$
Ba^{2+}	$1.6\cdot10^{-5}$	$2.2\cdot10^{-3}$	—	—	$1\cdot10^{-8}$	$1.4\cdot10^{-6}$	—	—	—	—
Hg^{2+}	—	—	$1.6\cdot10^{-44}$	$3.2\cdot10^{-42}$	—	—	—	—	—	—
Hg_2^{2+}	$9\cdot10^{-13}$	$3.6\cdot10^{-10}$	$1\cdot10^{-35}$	$4\cdot10^{-33}$	$1\cdot10^{-4}$	$4\cdot10^{-2}$	$1.1\cdot10^{-16}$	$4.4\cdot10^{-14}$	—	—
Sn^{2+}	$1\cdot10^{-16}$	$1.2\cdot10^{-14}$	—	—	—	—	—	—	—	—
Pb^{2+}	$1.5\cdot10^{-9}$	$3.1\cdot10^{-7}$	$7\cdot10^{-19}$	$1.4\cdot10^{-16}$	$1.3\cdot10^{-6}$	$2.7\cdot10^{-4}$	—	—	$1\cdot10^{-12}$	$2\cdot10^{-10}$
Fe^{2+}	$2.1\cdot10^{-7}$	$1.2\cdot10^{-5}$	$4\cdot10^{-9}$	$2.2\cdot10^{-7}$	—	—	—	—	—	—
Co^{2+}	$8\cdot10^{-9}$	$4.7\cdot10^{-7}$	$5\cdot10^{-12}$	$2.9\cdot10^{-10}$	—	—	—	—	—	—
Ni^{2+}	$1.4\cdot10^{-3}$	$8.1\cdot10^{-2}$	$3\cdot10^{-11}$	$1.7\cdot10^{-9}$	—	—	—	—	—	—

Note: Maximal concentrations of metal ions are shown in mole/liter and g/liter. All data are for 25° C and 1 atm.

1 g/liter. Most minerals possess such low solubilities (Table 12), making this rule very useful to geochemists.

The concentration of one ion in the given solution being known, concentrations of all other ions may be determined approximately by this rule. For example, it can be seen from the data of Table 12 that values of solubility products are very small for most sulfides. Let us assume that a given natural water contains 10^{-10} gram-ion/liter of S^{2-}, or, which is the same, $3.2 \cdot 10^{-9}$ g/liter of sulfur. We note that the solubility product for CuS is $8 \cdot 10^{-37}$, i.e.,

$$[Cu^{2+}][S^{2-}] = 8 \cdot 10^{-37}.$$

Substituting for S^{2-} its assumed content, 10^{-10}, we calculate the concentration of Cu at the point of equilibrium:

$$[Cu^{2+}] = \frac{8 \cdot 10^{-37}}{[S^{2-}]} = \frac{8 \cdot 10^{-37}}{10^{-10}} = 8 \cdot 10^{-27} \text{ g} \cdot \text{ion/liter.}$$

Changing from molar concentrations to grams, we get $5 \cdot 10^{-25}$ g per liter Cu — an amount not detectable by chemical analysis. Since natural waters ordinarily contain only 10^{-6} g/liter Cu, even minute amounts of sulfide ion will precipitate virtually all the Cu present. Low solubility products of other sulfides testify to the precipitating role of H_2S. Since surface or ground waters contain about 10^{-6} to 10^{-8} g/liter each of Cu, Zn, Ni, Co, Fe, and Hg, the precipitation of these metals with hydrogen sulfide should be practically complete in alkaline waters.

The solubilities of the most important compounds of rare metals — carbonates, sulfides, sulfates, chlorides, and phosphates — were calculated by Krauskopf* (Table 12). Knowing the solubility product of a given compound, the table gives at a glance the concentration of its metal in a saturated solution. Although the data are theoretical and do not take into account temperature fluctuations or the presence of other ions in natural waters, they are useful in geochemistry. For example, it may be calculated that sulfates of rare metals do not precipitate from natural waters in deserts and semideserts. Such waters contain only $1 \cdot 10^{-6}$ to $1 \cdot 10^{-5}$ g/liter of such metals, and it would take $2.7 \cdot 10^{-4}$ g/liter Pb^{2+} or $6.6 \cdot 10^{-3}$ g/liter Sr^{2+} to start the precipitation of sulfates.

*He used somewhat different data and a different method of calculation, and obtained somewhat different results.

Thus, the sulfide ion will precipitate these metals from natural waters — surficial or ground — while the sulfate ion will not. Anglesite can form only where the oxidation of lead ores produced higher concentrations of Pb^{2+}. On the other hand, lead sulfide can form wherever the sulfide ion is available. This view is confirmed by occurrences of galena and sphalerite in coal seams where we could hardly expect high concentrations of Pb and Zn in circulating waters. It is interesting to note that black marine clays are rich in sulfides of various metals but not in sulfates. Calculations show that ground waters containing $1 \cdot 10^{-6}$ g/liter PO_4^{3-} and the same amount of Pb^{2+} or Zn^{2+} will precipitate Pb but not Zn.

As it has been stated, exact forms of the occurrence of elements in natural water must be known before calculations involving solubility products can be made. All chemical calculations of this nature should be corrected for actual geochemical conditions.

IONIZATION POWER OF NATURAL WATERS

Interacting ions usually create an electric field in natural waters which reduces the effect of active mass. The quantity of a substance which actually participates in a given reaction is called the active concentration or activity, a. The factor by which the given concentration C should be multiplied to obtain the activity is known as the coefficient of activity, f. Thus, $a = fC$.

Since this coefficient is equal to 1 only for extremely dilute solutions, the activity equals the molar concentration.

The ionization power of a solution is a measure of the intensity of the electric field which conditions the size of the coefficient of activity. It is denoted by the letter μ.

$$\mu = \frac{C_1 Z_1^2 + C_2 Z_2^2 + \ldots + C_n Z_n^2}{2},$$

where the C's are ionic concentrations and the Z's are ionic valences.

Thus, the ionization power is equal to one-half of the sum of the products of the molar concentrations and the squares of the valence. It has been established that for solutions of equal ionization power, activity coefficients should be the same for all ions of the same charge (Table 13). This rule holds exactly for μ values up to 0.02, but can still be used for values up to 0.2. As

Table 13. Activity Coefficients (f) of Ions
in Solutions of Different Power

Ionization power of solution	Activity coefficients		
	Monovalent ions	Bivalent ions	Trivalent ions
0.00	1	1	1
0.001	0.96	0.87	0.72
0.002	0.95	0.81	0.63
0.005	0.92	0.72	0.48
0.01	0.89	0.63	0.35
0.02	0.87	0.57	0.28
0.05	0.81	0.43	0.15
0.1	0.76	0.34	0.084
0.2	0.70	0.24	0.041
0.5	0.62	0.15	0.014

an example of the calculation, let us take a natural water of the following composition (mg/liter):

Cl^-	HCO_3^-	SO_4^{2-}	Ca^{2+}	Mg^{2+}	Na^+	Total min-eralization
6.09	244.0	1560.00	504.00	95.16	84.41	2493.66

Recalculating to moles per liter, we get:

Cl^-	HCO_3^-	SO_4^{2-}	Ca^{2+}	Mg^{2+}	Na^+
$0.17 \cdot 10^{-3}$	$4.1 \cdot 10^{-3}$	$16.25 \cdot 10^{-3}$	$12.6 \cdot 10^{-3}$	$3.9 \cdot 10^{-3}$	$3.67 \cdot 10^{-3}$

Substituting into the formula, we have:

$$\mu = \frac{1}{2} \, [Cl^-] + [HCO_3^-] + 4[SO_4^{2-}] + 4[Ca^{2+}] + 4[Mg^{2+}] + [Na^+] = \frac{1}{2}(0.17 \cdot 10^{-3}$$
$$+ 4 \times 10^{-3} + 65 \cdot 10^{-3} + 50.4 \cdot 10^{-3} + 15.60 \cdot 10^{-3} + 3.6 \cdot 10^{-3}) = 0.069.$$

We find from Table 13 that activity coefficients are 0.81 for Na^+, Cl^-, and HCO_3^-, and 0.44 for Ca^{2+}, Mg^{2+}, and SO_4^{2-}.

Multiplying molar concentrations by these coefficients we obtain activities of ions present in the water:

Cl^-	HCO_3^-	SO_4^{2-}	Ca^{2+}	Mg^{2+}	Na^+
$0.14 \cdot 10^{-3}$	$3.32 \cdot 10^{-3}$	$7.15 \cdot 10^{-3}$	$5.54 \cdot 10^{-3}$	$1.71 \cdot 10^{-3}$	$3.10 \cdot 10^{-3}$

It will be noted that for bivalent ions active concentration is considerably smaller than actual (e. g., $7.15 \cdot 10^{-3}$ instead of $16.25 \cdot 10^{-3}$ for SO_4^{2-}). These differences are small for monovalent ions. Active concentrations may be taken as equal to actual concentrations for very weakly mineralized waters (up to 100 mg/liter). However, the ionization power of most natural waters exceeds 0.005 and their active concentrations are not equal to the actual concentrations.

An increase in the mineralization of natural water lowers the coefficient of activity and the active concentration of the ions. The solubility of difficultly soluble salts improves in the presence of some foreign salt. For instance, the solubility of $CaSO_4$ in water is 2 g/liter but in 0.1 M solution of NaCl it increases to 3.3 g/liter. This phenomenon is known in chemistry as the salinity effect. It should be taken into account during salinity studies in deserts and semideserts. It speeds up the migration of many elements.

With the concept of activity in mind, we can rewrite the law of mass action in the following manner:

$$\frac{a_R^k \cdot a_S^l \ldots a_Q^u}{a_B^n \cdot a_C^m \ldots a_D^z} = K,$$

where the a's denote the activities of the ions and molecules B, C, D, R, S, and Q taking part in the reaction and n, m, ...z, k, l, ..., u are their stoichiometric amounts.

CONCENTRATION OF ELEMENTS IN NATURAL WATERS AND ITS SIGNIFICANCE TO THE FORMATION OF MINERALS

It follows from the solubility product rule that, all other factors remaining the same, those elements form minerals which can concentrate in water. Rare anions become bonded to common cations: Ca^{2+}, Mg^{2+}, etc. Rare cations become tied to common anions: PO_4^{3-}, CO_3^{2-}, etc. A low abundance of an element produces its low concentrations in natural waters. No sediment forms since its solubility product was not reached. This is the reason for the commonness of sulfates and the scarcity of selenates within the supergene zone (the abundances of sulfur and selenium are 0.05% and $6 \cdot 10^{-5}\%$, respectively). While sulfates are found everywhere in deserts and semideserts, selenates form only in areas of selenium mineralization. There we find $CaSeO_4$, Na_2SeO_4, and $PbSeO_4$, which consist of common cations and rare anions, but not Li_2SeO_4, Rb_2SeO_4, $BeSeO_4$ (rare cation – rare anion). Similarly, a rare anion CrO_4^{2-}

Table 14. Average Numbers of Minerals Possible for One
Element of Each Decade (after Saukov, 1946)

Decade	Clarke, weight %	Average number of minerals possible for one element	Decade	Clarke, weight %	Average number of minerals possible for one element
I	> 10	729	VI	$10^{-3}-10^{-4}$	23
II	$1-10$	239	VII	$10^{-4}-10^{-5}$	28
III	$10°-10^{-1}$	139	VIII	$10^{-5}-10^{-6}$	23
IV	$10^{-1}-10^{-2}$	31	IX	$10^{-6}-10^{-7}$	2
V	$10^{-2}-10^{-3}$	28	X	$< 10^{-7}$	< 1

combines with Pb in zones of oxidation of sulfide ore deposits and with common cations Ca^{2+}, Mg^{2+}, and K^+. The formation of such minerals as Li_2CrO_4, Rb_2CrO_4, and $SrCrO_4$ is improbable. The significance of abundance is definite even though the ionic concentrations of elements in solution are conditioned by their chemical properties in addition to abundance. Low abundances of rare elements and their correspondingly low contents in natural waters are principal causes of the scarcity of their naturally occurring compounds.

The average number of minerals which could form from elements of different abundance were calculated by Saukov (Table 14).

Thus, the number of possible minerals is considerably smaller than the number of compounds made in laboratory. Although mineralogists and chemists strive to explain the presence of minerals analogous to artificial compounds, their absence is equally useful. For this purpose we are introducing the concept of the coefficient of mineral formation, M. It denotes the ratio of the number of minerals to the number of compounds of the same type.

Chapter 4

Analysis of Energy Involved
in Supergene Migration

Supergene processes involve, as do all other natural processes, the transformation of matter and energy. Any geochemical investigation should be concerned with both sides of a given process of migration.

The need for energy analysis was evident to the founders of geochemistry, V. I. Vernadskii and A. E. Fersman. In 1937, Fersman wrote:

> "It is quite clear to me that laws of leaching and the laws of weathering are among the most important objectives of modern geochemistry and should be verified by means of the analysis of the energy involved in those processes."

Earlier, opinions on the significance of energy were written by Polynov (1917, 1934), Van Hise (1904), Leith and Mead (1915), and many other investigators.

The study of chemical thermodynamics was well suited to the purposes of evaluating the conditions of mineral existence and constructing ranges of stability (Eh-pH diagrams, etc.). The stability ranges of polyvalent elements were calculated by Garrels, Krauskopf, and others, this being the way to calculate the modes of occurrence of chemical elements in water. Chemical thermodynamics is especially well suited for determinating the directions of reactions.

Fersman's foresight on the significance of energy analyses in geochemistry has been confirmed. The feasibility of applying chemical thermodynamics was confirmed and the need for it substantiated. An important feature of natural processes mentioned by Vernadskii has been proved: "Natural reactions take place in conditions which are in most cases unknown to chemists."

A further development of such ideas we find in the works of Korzhinsky, who held that "Geologists investigating natural processes are confronted with new problems which are unusual for laboratory workers."

The investigation of metasomatism led him toward the concept of "open systems" characterized by inert and mobile components.

Chemical reactions proceed in supergene zones under conditions different from laboratory conditions, making Korzhinsky's premise of great interest to those concerned with energy analyses.

Recently, the use of chemical thermodynamics has become popular in energy analyses. Papers by Garrels (1962), Ginzburg, Bughelskii, Letnikov, and others should be mentioned in this connection.

Comparatively low temperatures and pressures are the typical characteristics of the supergene zone which distinguish it from deeper parts of the earth's crust. Many supergene reactions proceed at the constant pressure of one atm. In soils, weathered crusts, or surface waters, temperatures are, as a rule, within twenty degrees above freezing during the warm time of the year. Such conditions favor the low velocities of many reactions and retard a number of processes. The continuous input of solar energy ensures the continuity of supergene processes at the earth's surface and imparts to them a characteristic cyclicity (which is cycloidal). The water cycle and the biologic cycle of the atoms should be mentioned as examples.

Energy is either absorbed or evolved during reactions in the supergene zone, such reactions being termed endoenergetic and exoenergetic, respectively.

Photosynthesis is the most important endoenergetic reaction. Great numbers of organic compounds are formed in living organisms as a result of this reaction. Most of these compounds are stable only during the life of the organism. On this basis, Vernadskii concluded that special thermodynamic conditions should exist within living organisms which are absent in "dead" nature. All organic compounds are enriched in energy.

The evaporation of water, another endoenergetic process, is part of the water cycle on the earth, and, therefore, is related to all the processes of atom migration. Other endoenergetic (endothermal) processes include the dissolution of substances, the melting of ice, and the mechanical disintegration of rocks.

Each of these processes finds its counterpart in the supergene zone, i.e., there exists another process during which the energy is evolved instead of being absorbed. Broadly speaking, the counterpart of photosynthesis is the transpiration of organisms, which is accompanied by the oxidation of organic matter and the liberation of the energy accumulated in that matter.

The spontaneous combustion in coal seams or in haystacks testifies to the evolution of considerable heat during the decomposition of organic matter. The geothermal gradient is greater in regions underlain by rocks rich in organic matter.

Heat is not the only form in which energy is liberated during the mineralization of organic matter. Chemical forms of energy are also expended when chemically active substances are formed, such as carbon dioxide. Forming everywhere during the decomposition of organic matter, and dissolving in natural waters, CO_2 increases the chemical activity in these waters. This results in the corrosion of limestones and other carbonatites, the weathering of silicates, etc. Humic acids, themselves the products of the decomposition of organic matter, are prominent agents in chemical weathering. They lower the pH of waters to 4, thus enhancing their aggressiveness. Highly active in this respect are inorganic compounds of sulfur, phosphorus, and other elements which form during the decomposition of organic substances.

Diametrically opposite to the evaporation and melting of water are the processes of condensation and freezing, during which heat is evolved. The cementation of rocks, crystallization from solutions, sorption, and many other processes are also accompanied by the evolution of heat. However, the geochemical significance of most such processes are minimal when compared with the decomposition of organic matter. No other process imparts so much chemical energy to natural waters.

Thus, the mutually opposed endoenergetic and exoenergetic processes are both interrelated and interdependent in the supergene zone. Running simultaneously within a given system, each pair may be considered as one single supergene process. Endoenergetic processes—e. g., photosynthesis—predominate at the earth's surface. It is at the earth's surface, in general, that atoms are charged with solar energy, which they eventually lose when buried in deeper portions of the supergene zone. However, both types of mutually opposed processes run contiguously wherever life exists.

As is well known, classical thermodynamics deals mainly with closed systems characterized by the lack of the exchange of matter (but not necessarily energy) with the surroundings. The basic formulas of that discipline are applicable only to reversible processes and to systems in equilibrium. On the other hand, all systems in the supergene zone are open systems exchanging both heat and matter with their surroundings.

The continued inflow of solar energy produces fluctuations in temperature and in the speed of migration of chemical elements, and generally disrupts the established chemical equilibria.

Still, chemical equilibria are possible in some parts of the supergene zone. The "mosaic of equilibria" concept developed by Korzhinsky (1962) is widely used. According to Korzhinsky, a system as a whole may be unbalanced, with different static processes operating in spots due to differences in temperature, pressure, or concentration; however, these parameters have constant values at each point of the system. It is only natural that a chemical equilibrium will be reached eventually between the existing phases and in accord with conditions existing at each point. Such a local equilibrium at any given point will be preserved even for the case where the process is not static and the parameters change at a rate slower than that at which equilibrium is attained. Thus, we come to the concept of mosaics of local equilibria within a system not in equilibrium.

Such mosaics should be typical in water-bearing horizons and in pore water in rocks. Conceivably, such mosaics could also develop in the weathered crust, in soils, and in surface waters. In fact, mosaics of equilibria probably exist in oozes of rivers, lakes, and seas (due to processes of diagenesis), in salt lakes, etc.

CHEMICAL AFFINITY

A question which is raised often during studies of supergene processes is whether chemical reactions are predetermined by the existing temperature, pressure, or concentration. An experimental solution of the question is seldom possible, while a theoretical analysis on the basis of thermodynamics is satisfactory. In chemistry, the problem is solved by determining chemical affinities,

i. e., the abilities of elements to form compounds with each other. For example, a considerable chemical affinity is observed between free atmospheric oxygen and pyrite when the latter is spontaneously oxidized. On the other hand, there is no affinity in supergene environments between native gold and oxygen, since these elements do not react. However, as will be shown later, the absence of chemical interaction does not always indicate the absence of affinity, since reactions might be retarded. Chemical affinity varies with pressure and temperature. Therefore, an affinity may be nil at the low temperatures of the supergene zone but may become pronounced at the elevated pressures and temperatures in hypogene environments. For example, oxides of iron and titanium do not interact under supergene conditions, their mutual affinity being nil in those conditions. They react, however, in magmatic conditions and form a number of minerals, such as ilmenite, $FeTiO_3$.

In the second half of the last century, M. Berthelot (1867) suggested that the heat of reaction could be used for the evaluation of chemical affinity. This premise has become known as the Berthelot-Thomson principle. Kotyukov (1963) restated this principle in a form very convenient for our purposes: "Of all reactions possible in a given system, that reaction will take place which is accompanied by the evolution of the greatest amount of heat."

This principle is particularly applicable to irreversible reactions occurring at low temperatures in condensate systems, i. e., to systems which do not contain gases.

Great heat is evolved during the oxidation of sulfides, which indicates their considerable affinity toward oxygen and their oxidizability. The oxidation of sulfur, hydrogen sulfide, or ferrous iron evolves a lesser amount of heat. Calculations show that the oxidation of gold should be accompanied by the absorption of heat; as we know, gold is a noble metal not oxidizable in the supergene zone.

The applicability of the Berthelot-Thomson principle was studied by V. Nernst, who noted that the principle is a rule and not a law of Nature, since it cannot be applied satisfactorily in all cases.

Nernst also drew attention to the fact that chemical affinities and heats of reaction often parallel each other, particularly for solid substances.

The applicability of the Berthelot-Thomson principle has been discussed by Vernadskii and Kurbatov and by Fersman. There are processes operating within the supergene zone which do not follow this rule. The directions of many reactions are determined by the concentrations of the reacting substances. In other words, such reactions are reversible and can proceed in either direction. Examples of such reactions are the precipitation and dissolution of salts:

$$\underline{CaCO_3} + H_2O + CO_2 \rightleftharpoons Ca(HCO_3)_2,$$
$$\downarrow solid$$

$$Ba^{2+} + SO_4^{2-} \rightleftharpoons \underline{BaSO_4}.$$
$$\downarrow solid$$

A number of endothermal reactions proceed at random. This fact has led chemists to conclude that the reaction's heat does not precisely express chemical affinity. In fact, it often does not even characterize the affinity.

The correct interpretation appeared later in the works of Gibbs (1878) and Helmholtz (1884). Their ideas were translated into mathematical equations by Van't Hoff (1885). The latter showed that chemical affinity should be measured by the maximum work performed by the forces* during an isothermal reversible process. That some work is actually performed during a chemical reaction can be readily seen, for example, in electric batteries. The electric energy so obtained may be transformed into some other form of energy or into mechanical work. The reactions in electric batteries are very nearly reversible, and the batteries can be used for measuring the work done by the forces of chemical affinity.

In a given reversible process, the work done by the forces of chemical affinity depends upon the nature of the reaction itself, the temperature, and the initial concentrations. At constant pressure and temperature, the work A is the negative value of the isobar potential ΔZ:

$$A = -\Delta Z.$$

As shown in Chapter 3, we can express a reversible reaction proceeding at constant pressure and temperature by:

$$nB + mC + \ldots \rightleftharpoons kR + lS + \ldots.$$

This means that n moles of B and m moles of C have taken part in the reaction producing k moles of S, etc. The quantities of B, C, R,

*The choice of the terms "force" and "chemical affinity" is unfortunate, but has been established in the literature. Rakowsky pointed this out in his Manual of Physical Chemistry.

S, etc., should be so large that an expenditure of n moles of B and m moles of C is negligible enough not to affect either the composition or the existence conditions of the given system. Chemical equilibrium will be attained some time after the start of the reaction, i. e., the velocities of the direct and the reverse reactions will equalize. Let us denote active concentrations existing during the equilibrium as a_B, a_C, a_R, etc.

Then, according to chemical thermodynamics, the increment of the isobar potential for this reaction will be:

$$\Delta Z = RT \ln \left[\frac{a_R'^k \cdot a_S'^l \cdots}{a_B'^n \cdot ac_{...}'^m} \right] - RT \ln \left[\frac{a_R^k \cdot a_S^l \cdots}{a_B^n \cdot a_{C...}^m} \right].$$

The first term on the right in this equation represents the active concentration at the start of the reaction and before equilibrium (a'); the second term represents the same concentrations after equilibrium has been reached (a).

Should the value of ΔZ be negative, the reaction could proceed at random in either direction. If the value is positive, the reaction would be impossible. It follows from the foregoing that the reaction must inevitably take place once the sign of ΔZ is negative. Thermodynamics does not verify that. Therefore, the negative isobar potential only shows the direction of the reaction if the latter is possible at all.

The second term on the right in the above equation is the previously discussed equilibrium constant for a chemical reaction proceeding at a constant pressure, K_p. Substituting this into the equation for ΔZ, we have:

$$\Delta Z = RT \ln \frac{a_R'^k \cdot a_S'^l \cdots}{a_B'^n \cdot a_C'^m \cdots} - RT \ln K_p,$$

where R is the universal gas constant, 4987 cal/deg; and T is the absolute temperature.

Assuming that the activities of B, C, R, S, . . . equal unity, the first term on the right becomes zero, since $\ln 1 = 0$. The equation is simplified.

In such cases the value of the isobar potential at 25° C (298°·K) is called the standard potential, $\Delta Z°$. Substituting the appropriate values for R and T and transforming from natural to decimal logarithms, we have:

$$\Delta Z° = -1.364 \log K_p.$$

Table 15.   Standard Isobar Potentials of Ions
(after Latimer, 1954)

Cations

Li^+	-70.22	Cr^{3+}	-51.5
Na^+	-62.589	$Cr(OH)^{2+}$	-103.0
K^+	-67.46	Mn^{2+}	-54.4
Rb^+	-67.45	Mn^{3+}	(-19.6)
Cs^+	-67.41	Fe^{2+}	-20.30
Be^{2+}	-85.2	Fe^{3+}	-2.53
Be_2O^{2+}	-218	$Fe(OH)^{2+}$	-55.91
Mg^{2+}	-108.99	$Fe(OH)_2^+$	-106.2
Ca^{2+}	-132.18	Ce^{3+}	-171.75
Sr^{2+}	-133.2	$Ce(OH)^{3+}$	-188.9
Ba^{2+}	-134.0	$Ce(OH)_2^{2+}$	-245.2
Al^{3+}	-115.0	Co^{2+}	-12.8
Sc^{3+}	-143.7	Co^{3+}	28.9
Y^{3+}	-164.1	Ni^{2+}	-11.53
La^{3+}	-174.5	Cu^+	12.0
Th^{4+}	-175.2	Cu^{2+}	15.53
U^{4+}	-138.4	$Cu(NH_3)^+$	-2.8
UO_2^+	-237.6	$Cu(NH_3)_2^+$	-15.6
UO_2^{2+}	-236.4	$Cu(NH_3)_4^{2+}$	-40.8
Zr^{4+}	-141	Ag^+	18.430
ZrO^{2+}	-200.9	Au^+	39.0
V^{3+}	-60.6	Au^{3+}	103.6
$V(OH)^{2+}$	-112.8	Zn^{2+}	-35.184
$V(OH)_4^+$	-256	$Zn(NH_3)_4^{2+}$	-73.5
$Zn(OH)^+$	-78.8	Tl^{3+}	$+50.0$
Cd^{2+}	-18.58	In^{3+}	-23.7
$Cd(NH_3)_4^{2+}$	-53.73	$In(OH)^{2+}$	-55.5
Hg^{2+}	39.38	Sn^{2+}	-6.275
Hg_2^{2+}	36.35	Sn^{4+}	0.65
Ga^{3+}	-36.6	Pb^{2+}	-5.81
$GaOH^{2+}$	-89.8	Pb^{4+}	72.3
$Ga(OH)_2^+$	-142.1	NH_4^+	-19.00
Tl^+	-7.755		

Anions

BeO_2^{2-}	-155.3	CrO_4^{2-}	-176.1
$Be_2O_3^{2-}$	-298	MoO_4^{2-}	-218.8
BO_2^-	-169.6	WO_4^{2-}	-220
$H_2BO_3^-$	-217.6	$CuCl_2^-$	-57.9
$B_4O_7^{2-}$	-616.0	$Cu(CO_3)_2^{2-}$	-250.5
AlO_2^-	-204.7	$AuCl_4^-$	-56.2
$H_2AlO_3^-$	-255.2	ZnO_2^{2-}	-93.03
$UO_2(CO_3)_3^{4-}$	-640.0	HgS_2^{2-}	11.6
$UO_2(CO_3)_2(H_2O)^{2-}$	-622.0	$HgCl_4^{2-}$	-107.7
$HZrO_3^-$	-287.7	$HgBr_4^{2-}$	-88.0

Table   15   (continued)

Anions (continued)

VO_4^-	-203.9	GaO_3^{2-}	-148
$HV_6O_{17}^{3-}$	-1132	$H_2GaO_3^-$	-178
$H_2V_6O_{17}^{2-}$	-1135	HCO_3^-	-140.31
$V_4O_9^{2-}$	-665.3	CO_3^{2-}	-126.22
$H_2V_{10}O_{28}^{4-}$	-1875.2	$H_3SiO_4^-$	-286.8
$HV_{10}O_{28}^{5-}$	-1875.3	$HGeO_3^-$	-170
$V_{10}O_{28}^{6-}$	-1862.4	GeO_3^{2-}	-152.8
$Cr_2O_7^{2-}$	-315.4	$HSnO_2^-$	-98
$HCrO_4^-$	-184.9	HS^-	3.01
$Sn(OH)_6^{2-}$	-310.5	HSO_3^-	-126.0
$HPbO_2^-$	-81.0	HSO_4^-	-179.94
NO_2^-	-8.25	Se^{2-}	42.6
NO_3^-	-26.43	SeO_3^{2-}	-89.33
PO_4^{3-}	-245.1	SeO_4^{2-}	-105.42
HPO_3^{2-}	-194.0	HSe^-	23.5
HPO_4^{2-}	-261.5	$HSeO_3^-$	-98.3
$H_2PO_4^-$	-271.3	$HSeO_4^-$	-108.2
AsO_2^-	-83.7	Te^{2-}	52.7
AsO_4^{3-}	-152	Te_2^{2-}	38.75
$HAsO_4^{2-}$	-169	HTe^-	37.7
$H_2AsO_4^-$	-178.9	TeO_3^{2-}	-108.0
SbO_2^-	-82.5	F^-	-66.08
SbS_3^{2-}	(-32)	HF_2^-	-137.5
S^{2-}	22.1	Cl^-	-31.350
S_2^{2-}	21.8	Br^-	-24.574
SO_3^{2-}	-116.1	BrO_3^-	(-5.0)
SO_4^{2-}	-177.34	I^-	-12.35
$S_2O_3^{2-}$	-124.0	I_3^-	-12.31
$S_2O_4^{2-}$	-143.4	IO_3^-	-32.25
$S_2O_5^{2-}$	-189		
$S_2O_6^{2-}$	-231		
$S_2O_8^{2-}$	-262		
$S_3O_6^{2-}$	-229		
$S_4O_6^{2-}$	-244.3		

Note: Values given in parentheses have been interpolated.

The generalized formula expressing both the maximum work performed during the reaction and the isobar potential itself is:

$$-A = \Delta Z = RT \ln \frac{a_R'^k \cdot a_S'^l \cdots}{a_B'^n \cdot a_C'^m \cdots} + \Delta Z^\circ.$$

Standard isobar potentials are generally given in handbooks on physical chemistry and thermodynamics for many inorganic compounds. Among such books, those by Latimer (1954) and M. Kh. and M. L. Karapet'yants (1961) are particularly valuable. Standard potentials for some 340 minerals have been collected by Letnikov (1964) who himself had calculated many.

Taking a value for $\Delta Z°$ from a handbook and knowing the starting active concentrations of substances, it is possible to calculate the maximum work of the reaction (its isobar potential) by means of the above formula. At the same time, the direction of the reaction and the force of chemical affinity between the reacting substances are evaluated. Some standard isobar potentials for ions and compounds are given in Tables 15 and 16. As an example, let us consider the precipitation reaction of calcium fluoride:

$$\frac{Ca^{2+} + 2F^-}{\text{dil. sol.}} \rightleftarrows \frac{CaF_2}{\text{solid}}.$$

We find from Latimer's tables:

$$\Delta Z° \, Ca^{2+} = -132.18$$
$$\Delta Z° \, F^- = -66.08$$
$$\Delta Z° \, Ca F_{\text{solid}} = -277.7.$$

Then the standard isobar potential of the reaction of CaF_2 precipitation is calculated:

$$\Delta Z°_{\text{reaction}} = \Delta Z° CaF_2 - \Delta Z° Ca^{2+} - 2\Delta Z° F^- =$$
$$= -277.7 - (-132.18) - 2(-66.08) = -13.36.$$

The result indicates that calcium fluoride precipitates when the active concentrations of calcium and fluorine are equal to one gram-ion/liter (at 25° C). This is true for standard conditions only.

Let us consider now the actual concentrations characteristic of the supergene zone. Let us assume the active concentrations of Ca^{2+} ion and of F^- anion to be 10^{-2} and 10^{-3} mole/liter, respectively (0.4 and $9 \cdot 10^{-3}$ g/liter). Such contents are typical of ground waters of deserts and semideserts, which usually run higher in fluorine. The standard isobar potential is again calculated from the formula:

$$\Delta Z = RT \ln \frac{[CaF_2]}{[Ca^{2+}][F^-]^2} + \Delta Z° = 1.364 \, \log \frac{[CaF_2]}{[Ca^{2+}][F^-]^2} + \Delta Z°.$$

Since the concentration of CaF_2 occurring in the solid phase may be taken as unity, the equation is rewritten for 25° C:

$$\Delta Z = 1.364 \log [Ca^{2+}]^{-1}[F^-]^{-2} - 13.36.$$

Substituting, we get:

$$\Delta Z = 1.364 \log 10^2 \cdot 10^6 - 13.36 = 8 \cdot 1.364 - 13.36 = -2.45.$$

Because of the negative sign of the result, the reaction has to proceed from left to right, i. e., toward the precipitation of CaF_2.

In the case of weakly mineralized waters in regions with humid climates and an active circulation of water, conditions for the precipitation of calcium fluoride will be different. Let us assume the following active concentrations: 10^{-3} mole/liter or 0.04 g/liter for Ca^{2+} and 10^{-5} mole/liter or $9 \cdot 10^{-5}$ g/liter for F^-. Then,

$$\Delta Z = 1.364 \cdot \log [Ca^{2+}]^{-1}[F^-]^{-2} - 13.36 = 1.364 \log 10^3 \cdot 10^{10} -$$

$$- 13.36 = 13 \cdot 1.364 - 13.36 = 17.73 - 13.36 = +4.37.$$

No precipitation will take place under such conditions. In other words, the precipitation of calcium fluoride and the formation of supergene fluorite are impossible in weakly mineralized ground waters in humid climates. To the contrary, such waters dissolve fluorite.*

This simple example shows that to establish the direction of the reaction we should use the actual isobar potentials, ΔZ, and not the standard ones, $\Delta Z°$.

It should be noted that in many instances, analogous results might be obtained by using the solubility product because of its rather simple relationship to the isobar potentials, as will be explained later. Garrels (1960) has furnished a sample of the calculation of standard isobar potentials of reaction for solid substances:

$$PbCO_3 + CaSO_4 \rightleftarrows PbSO_4 + CaCO_3,$$

$$\Delta Z°_{reaction} = \Delta Z° PbSO_4 + \Delta Z° CaCO_3 - \Delta Z° CaSO_4 - \Delta Z° PbCO_3 =$$

$$= (-193.89) + (-269.78) - (-149.7) - (-315.56) = +1.59 \text{ cal}.$$

It follows that at 25°C and 1 atm cerussite and anhydrite are more stable than anglesite and calcite.

---

*Calculations made in this text hold only for the simplest, "ideal" conditions. For greater accuracy, the modes of occurrence of the elements in water should be ascertained— colloids, complex ions, etc. Their active concentrations should be used at a given temperature. Nevertheless, our example is very close to reality, since conditions for the precipitation of fluorite are more favorable in arid than humid climates.

Table 16. Standard Isobar Potentials of Compounds*

Chemical formula	Crystal- line	Dis- solved	Chemical formula	Crystal- line	Dis- solved
	Arsenates				
KH_2AsO_4	—237.0	—	$Sr_3(AsO_4)_2$	—770.1	—
$Mg(AsO_4)_2$	—679.3	—	H_3AsO_3	—	—152.94
$CaHAsO_4$ H_2O	—363	—	H_3AsO_4	—	—183.8
	Tungstates				
$FeWO_4$	—250.4	—	$BaWO_4$	—373.6	—
$CaWO_4$	—368.7	—	Ag_2WO_4	—206.0	—
$SrWO_4$	—366.5	—			
	Carbonates				
Li_2CO_3	—270.66	—266.66	$MnCO_3$	—195.4	—
Na_2CO_3	—250.4	—251.4	$MnCO_3$ (Precip)	—194.3	—
$NaHCO_3$	—203.6	—202.89	$FeCO_3$ Siderite	—161.06	—
K_2CO_3	—255.5	—261.2	$CoCO_3$	—155.57	—
Rb_2CO_3	—249.3	—	UO_2CO_3	—377.0	—
Cs_2CO_3	—243.6	—261.0			
$CsHCO_3$	—198.8	—			
H_2CO_3	—	—149.0	$NiCO_3$	—147.0	—
			$CuCO_3$	—123.8	—
			Ag_2CO_3	—104.48	—
$MgCO_3$	—246	—	$ZnCO_3$	—174.8	—
$CaCO_3$ Calcite	—269.78	—	$CdCO_3$	—160.2	—
$CaCO_3$ Aragonite	—269.53	—	Hg_2CO_3	—105.8	—
$SrCO_3$	—271.9	—	$PbCO_3$	—149.7	—
$Sr(HCO_3)_2$	—295	—	$PbOPbCO_3$	—195.6	—
$BaCO_3$	—272.2	—	$2PbOPbCO_3$	—242	—
			$Pb_3(OH)_2(CO_3)_2$	—409.1	—
	Nitrates				
$LiNO_3$	— 93.1	— 96.63	$Mg(NO_3)_2$	—140.63	—
$NaNO_3$	— 87.45	— 89.00	$Ca(NO_3)_2$	—177.34	—
KNO_3	— 93.96	—	$Sr(NO_3)_2$	—186	—
$RbNO_3$	— 93.3	— 93.86	$Ba(NO_3)_2$	—190.0	—
$CsNO_3$	— 94	— 93.82	HNO_3	— 19.1	— 26.43
	Oxides				
MgO	—136.13	—	MoO_2	—117.3	—
MgO (Finely ground)	—135.31	—	MoO_3	—161.95	—
CaO	—144.4	—	WO_3	—182.47	—
SrO	—133.8	—	Mn_2O_3	—212.3	—
BaO	—126.3	—	FeO	— 58.4	—
Al_2O_3 α —Corundum	—376.77	—	Fe_2O_3	—177.1	—
$Al_2O_3 \cdot H_2O$	—435	—	Fe_3O_4	—242.4	—
$Al_2O_3 \cdot 3H_2O$	—547.9	—	Co_3O_4	—179.4	—
La_2O_3	—426.9	—	CuO	— 30.4	—
ThO_2	—278.4	—	Cu_2O	— 34.98	—

*Table 16 after Latimer, 1954.

Table 16 (continued)

Chemical formula	Crystal-line	Dis-solved	Chemical formula	Crystal-line	Dis-solved
		Oxides (	continued)		
TiO_2 (Rutile)	−203.8	—	Ag_2O	− 2.586	—
TiO_2 (Hydrate)	−196.3	—	AgO	2.6	—
$FeTiO_3$	−268.9	—	HgO (Red)	− 13.990	—
ZrO_2	−244.4	—	HgO (Yellow)	− 13.959	—
HfO_2	−258	—	CO_2 (Gas)	−94.2598	− 92.31
V_2O_3	−271	—	SiO_2	−192.4	—
V_2O_4	−318	—	SiO_2 Cristobalite	−192.1	—
V_2O_5	−344	—	SiO_2 Tridymite	−191.9	—
Nb_2O_4	−362.4	—	SiO_2 Glass	−190.9	—
Nb_2O_5	−432	—	GeO_2 (Precip)	−127	—
Ta_2O_5	−470.6	—	SnO_2	−124.2	—
Cr_2O_3	−250.2	—	$As_2O_5 \cdot 4H_2O$	(−411.1)	—
PbO (Red)	− 45.25	—	Sb_2O_5	−200.5	—
PbO (Yellow)	− 45.05	—	Bi_2O_3	−118.7	—
PbO_2	− 52.34	—	SO_2 (Gas)	− 71.79	—
Pb_3O_4	−147.6	—	SO_3 (Gas)	− 88.52	—
NO_2 (Gas)	− 12.39	—	SeO_2	− 41.5	—
As_2O_5	−184.6	—			

Hydroxide

Chemical formula	Crystal-line	Dis-solved	Chemical formula	Crystal-line	Dis-solved
$Be(OH)_2$ α	—	−196.2	$Mn(OH)_3$	−181	—
$Be(OH)_2$ β	—	−195.5	$Fe(OH)_2$	−115.57	—
$Mg(OH)_2$	−199.27	—	$Fe(OH)_3$	−166.0	—
$Ca(OH)_2$	−214.33	—	$Ce(OH)_3$	−311.63	—
$Sr(OH)_2$	−207.8	—	$Co(OH)_2$	−109.0	—
$Ba(OH)_2$	−204.7	—	$Co(OH)_3$	−142.6	—
$Ba(OH)_2 \cdot 8H_2O$	−666.8	—	$Ni(OH)_2$	−108.3	—
$Al(OH)_3$ (Amorphous)	—	−271.9	$Pd(OH)_2$	− 72	—
$Sc(OH)_3$	—	−(293.5)	$Cu(OH)_2$	− 85.3	—
$Y(OH)_3$	—	−307.1	$Au(OH)_3$	− 69.3	− 61.8
$La(OH)_3$	—	−313.2	$Zn(OH)_2$	−132.6	—
$Th(OH)_4$	−379	—	$Cd(OH)_2$	−112.46	—
$U(OH)_3$	(−263.2)	—	$Hg(OH)_2$	—	− 65.70
$U(OH)_4$	(−351.6)	—	$Ga(OH)_3$	−199	—
$TiO(OH)_2$	−253.0	—	$Tl(OH)$	− 45.5	—
$ZrO(OH)_2$	−311.5	—	$Tl(OH)_3$	−123.0	—
$Zr(OH)_4$	−370	—	$In(OH)_3$	−182	—
$Hf(OH)_4$	(−325.5)	—	$Sn(OH)_2$	−117.6	—
$V(OH)_3$ (Precip)	−218.0	—	$Sn(OH)_4$	−227.5	—
$VO(OH)_2$ (Precip)	−213,6	—	$Pb(OH)_2$	−100.6	—
$Cr(OH)_3$	−215.3	—	NH_4OH	—	− 63.05
$Cr(OH)_3$ (Hydrate)	−205.5	—	$Bi(OH)_3$	−137	—
$Mn(OH)_2$ (Precip)	−146.9	—			

Selenates and Selenites

Chemical formula	Crystal-line	Dis-solved	Chemical formula	Crystal-line	Dis-solved
$PbSeO_4$	−122.0	—	H_2SeO_4	—	−105.42
Ag_2SeO_4	− 68.5	—	H_2SeO_3	—	−101.8

Selenides

Chemical formula	Crystal-line	Dis-solved	Chemical formula	Crystal-line	Dis-solved
$FeSe$	− 13.9	—	$ZnSe$	− 34.7	—
$PbSe$	− 15.4	—	$CuSe$	− 7.9	—
Tl_2Se	− 19.8	—	H_2Se (Gas)	17.0	18,4

Table 16 (continued)

Chemical formula	Crystal-line	Dis-solved	Chemical formula	Crystal-line	Dis-solved

Silicates

Chemical formula	Crystal-line	Dis-solved	Chemical formula	Crystal-line	Dis-solved
Li_2SiO_3 Glass	−356.1	—	$BaSiO_3$	−338.7	—
Na_2SiO_3	−341	—	Ba_2SiO_4	−470.6	—
$KAlSi_3O_8$	−856.0	—	Al_2SiO_5 Andalusite	−607.8	—
$KAlSi_3O_{10}(OH)_2$	−1298	—	Al_2SiO_5 Kyanite	−607.0	—
H_2SiO_3	−244.5	—	Al_2SiO_5 Sillimanite	−615.0	—
H_4SiO_4	—	−300.2	$Al_2Si_2O_5(OH)_4$ Ka-	−883.0	—
$CaSiO_3$ Pseudo-	−357.4	—	olinite		
wollastonite			$MnSiO_3$	−283.3	—
$CaSiO_3$ Wollastonite	−358.2	—	$FeSiO_3$	−319.8	—
Ca_2SiO_4 (β)	−512.7	—	$ZnSiO_3$	−274.8	—
Ca_2SiO_4 (γ)	−513.7	—	$PbSiO_3$	−239.0	—
$SrSiO_3$	−350.8	—	$PbSiO_4$	−285.7	—
$SrSiO_4$	−495.7	—			

Sulfates

Chemical formula	Crystal-line	Dis-solved	Chemical formula	Crystal-line	Dis-solved
Li_2SO_4	−316.6	−317.78	$BeSO_4$	−260.2	—
Na_2SO_4	−302.78	−302.52	$NiSO_4$	−184.9	—
K_2SO_4	−314.62	—	$NiSO_4 \cdot 6H_2O$	−531.0	—
Rb_2SO_4	−312.8	−312.24	$CuSO_4$	−158.2	—
Cs_2SO_4	−310.7	−312.16	Cu_2SO_4	−156	—
$MgSO_4$	−280.5	—	$CuSO_4 \cdot H_2O$	−219.2	—
$SrSO_4$	−318.9	—	$CuSO_4 \cdot 3H_2O$	−334.6	—
$BaSO_4$	−323.4	—	$CuSO_4 \cdot 5H_2O$	−449.3	—
$RaSO_4$	−326.0	—	$Cu_4(OH)_6SO_4$	−434.62	—
$CaSO_4$ Anhydrite	−315.56	—	Ag_2SO_4	−147.17	—
$CaSO_4$ (α)	−313.52	—	$Cu_4(OH)_6SO_4 \cdot$	−505.5	—
$CaSO_4$ (β)	−312.46	—	$\cdot 1,3\ H_2O$		
$CaSO_4 \cdot \frac{1}{2}H_2O$ (α)	−343.02	—	$Cu_3(OH)_4SO_4$	−345.5	—
$CaSO_4 \cdot \frac{1}{2}H_2O$ (β)	−342.78	—	$ZnSO_4$	−208.31	—
$CaSO_4 \cdot 2H_2O$	−429.19	—	$ZnSO_4 \cdot H_2O$	−269.9	—
$Al_2(SO_4)_3$	−738.99	—	$ZnSO_4 \cdot 6H_2O$	−555.0	—
$Al_2(SO_4)_3 \cdot 6H_2O$	−1105.14	—	$ZnSO_4 \cdot 7H_2O$	−611.9	—
$Y_2(SO_4)_3 \cdot 8H_2O$	−1327.3	—	$CdSO_4$	−195.99	—
$CsAl(SO_4)_2\ 12H_2O$	−1218.5	—	$Zr(SO_4)_2$	−538.9	—
H_2SO_3	—	−128.59	UO_2SO_4	—	−413.7
H_2SO_4	—	−177.34	$UO_2SO_4 \cdot 3H_2O$	−586.0	—
$MnSO_4$	−228.48	—	$HgSO_4$	−141.0	—
$Mn_2(SO_4)_3$	−580.9	—	Hg_2SO_4	−149.12	—
$Fe(SO_4)$	−198.3	—	$In_2(SO_4)_3$	−613.4	—
$FeSO_4 \cdot 7H_2O$	−597	—	Tl_2SO_4	−196.8	—
$CoSO_4$	−180.1	—	$Sn(SO_4)_2$	−346.8	—
$CdSO_4 \cdot H_2O$	−254.84	—	$PbSO_4$	−193.89	—
$CdSO_4 \cdot \frac{3}{8}\ H_2O$	−349.63	—	$PbSO_4PbO$	−258.9	—

Sulfides

Chemical formula	Crystal-line	Dis-solved	Chemical formula	Crystal-line	Dis-solved
MoS_2	− 53.8	—	Ag_2S (β)	− 9.36	—
MoS_3	− 57.6	—	ZnS Sphalerite	− 47.4	—
WS_2	− 46.2	—	ZnS Wurtzite	− 44.2	—
MnS (Green)	− 49.9	—	ZnS (Precip)	− 43.2	—
MnS (Precip)	− 53.3	—	CdS	− 33.6	—
FeS (α)	− 23.32	—	HgS Cinnabar	− 11.67	—

Table 16 (continued)

Chemical formula	Crystal-line	Dis-solved	Chemical formula	Crystal-line	Dis-solved
Sulfides (continued)					
FeS$_2$ Pyrite	— 39.84	—	HgS (Black)	— 11.05	—
CoS	— 19.8	—	Hg$_2$S	— 1.6	—
NiS (α)	— 17.7	—	Tl$_2$S	— 21	—
NiS (γ)	— 27.3	—	SnS	— 19.7	—
CuS	— 11.7	—	PbS	— 22.15	—
Cu$_2$S	— 20.6	—	H$_2$S (Gas)	— 7.892	— 6.54
Ag$_2$S (ortho-rhomb α)	— 9.62	—			
Dichromates					
KCr$_2$O$_4$	—310.5	—	Ag$_2$CrO$_4$	—154.7	—
CaCrO$_4$	—312.8	—	Hg$_2$CrO$_4$	—155.75	—
SrCrO$_4$	—315.3	—	PbCrO$_4$	—203.6	—
BaCrO$_4$	—324.0	—			
Phosphates					
Na$_3$PO$_4$	—430.6	—	Ca(H$_2$PO$_4$)$_2$(Precip)	—672	—
KH$_2$PO$_4$	—339.2	—	Sr$_3$(PO$_4$)$_2$	—932.1	—
H$_3$PO$_4$	—	—274.2	SrHPO$_4$	—399.7	—
H$_3$PO$_3$	—	—204.8	Ba$_3$(PO$_4$)$_2$	—944.4	—
Mg$_3$(PO$_4$)$_2$	—904	—	Mn$_3$(PO$_4$)$_2$ (Precip)	—683	—
Ca$_3$(PO$_4$)$_2$ (α)	—929.7	—	FePO$_4$	—272	—
Ca$_3$(PO$_4$)$_2$ (β)	—932.0	—	Pb$_3$(PO$_4$)$_2$	—581.4	—
CaHPO$_4$	—401.5	—	PbHPO$_3$	—208 3	—
CaHPO$_4 \cdot$ 2H$_2$O	—514.6	—			
Fluorides					
CaF$_2$	—277.7	—	CdF$_2$	—154.8	—
SrF$_2$	—277.8	—	PbF$_4$	—178.1	—
BaF$_2$	—272.5	—	PbF$_2$	—148.1	—
HF (Gas)	— 64.7	— 70,41	SbF$_3$	—199.8	—
BF$_3$ (Gas)	—261.3	—	SeF$_6$ (Gas)	—222.0	—
MnF$_2$	—179	—	TeF$_6$ (Gas)	—292.0	—
Chlorides					
AlCl$_3$	—152.2	—	SrCl$_2$	—186.7	—
FeCl$_2$	— 72.2	—	BaCl$_2$	—193.8	—
FeCl$_3$	— 80.4	—	BaCl$_2$. H$_2$O	—253.1	—
CoCl$_2$	— 65.5	—	BaCl$_2$. 2H$_2$O	—309.8	—
NiCl$_2$	— 65.1	—	CuCl	— 28.2	—
CuCl$_2$	— 42	—	AgCl	— 26.224	—
NaCl	— 91.785	— 99.939	ZnCl$_2$	— 88.255	—
KCl	— 97.592	— 98.816	CdCl$_2$	— 81.88	— 84.3
RbCl	— 96.8	— 98.8	HgCl$_2$	— 44.4	—
CsCl	— 96.6	— 88.76	SnCl$_2$	— 72.2	—
BeCl$_2$	—111.8	—	SnCl$_4$	—113.3	—
MgCl$_2$	—141.57	—	PbCl$_2$	— 75.04	—
CaCl$_2$	—179.3	—	HCl (Gas)	— 22.769	— 31.35

Isobar potentials characterize reactions proceeding at a constant pressure. Should a reaction proceed at a constant volume, its maximum work will be measured by another function, the isochore potential, ΔF, which is also known as the free energy. For isothermal and reversible reactions, the value of ΔF should be determined from an equation analogous to the one above:

$$-A = \Delta F = RT \ln \frac{a_R^{'k} \cdot a_S^{'l} \dots}{a_B^{'n} \cdot a_C^{'m} \dots} - RT \ln K_c,$$

where K_c is an equilibrium constant of the reaction proceeding at a constant volume.

Values of ΔZ and ΔF are nearly equal to each other for reactions taking place in condensed systems, i.e., in the absence of a gas phase.

Thus, the very existence of a reaction at given conditions is determined by the sign of its isobar potential. If the sign is negative, the reaction can proceed in the given direction. If the sign is positive, the given reaction is not feasible, but the reverse reaction is possible. If the isobar potential is equal to zero, the system is in equilibrium.

The negative sign of a ΔZ function does not always indicate the actual existence of the reaction. The reaction might be retarded or postponed by various obstacles. As a result, new systems with false, retarded equilibria may arise. This often happens in supergene environments. This retarding effect may be overcome by removing some specific energy barrier, such as the low temperature that caused it.

For example, the oxidation of carbonaceous substances by free atmospheric oxygen is a reaction producing a high thermal effect. The corresponding chemical affinity, i.e., the maximal work done during the reaction, is unquestionably large. However, low temperatures, which prevail in supergene zones, retard this oxidation so much that coal outcrops remain unaltered for a long time. The environmental temperature should be raised by several hundred degrees to start the energy reaction.

It does not follow that chemical reactions cannot proceed within the supergene zones unless accompanied by an increase in isobar potential. Such reactions are entirely possible, but will not start

Table 17. Standard Thermal Effects of Reactions at Constant
Pressure, $\Delta H°$, and Standard Isobar Potentials, $\Delta Z°$

Reaction	Energy factors	
	$\Delta H°$	$\Delta Z°$
$Cu^{2+} + S^{2-} \rightleftharpoons CuS$	—35.55	—49.33
$Zn^{2+} + S^{2-} \rightleftharpoons ZnS$	—20.63	—34.32
$Ba^{2+} + SO_4^{2-} \rightleftharpoons BaSO_4$	—4.63	—12.06
$Ca^{2+} + CO_3^{2-} \rightleftharpoons CaCO_3$	+2.95	—11.38
$Mg^{2+} + CO_3^{2-} \rightleftharpoons MgCO_3$	+6.04	—10.79
$Fe^{2+} + CO_3^{2-} \rightleftharpoons FeCO_3$	+4.63	—14.54
$Zn^{2+} + CO_3^{2-} \rightleftharpoons ZnCO_3$	+3.86	—13.4
$Ca^{2+} + 2 F^- \rightleftharpoons CaF_2$	—3.21	—13.36
$Pb^{2+} + CO_3^{2-} \rightleftharpoons PbCO_3$	—6.06	—17.67
$Fe^{2+} + Se^{2-} \rightleftharpoons FeSe$	—23.20	—36.2
$Cu^{2+} + Se^{2-} \rightleftharpoons CuSe$	—52.29	—66.03
$3Ca^{2+} + 2 PO_4^{3-} \rightleftharpoons Ca_3(PO_4)_2$	+16.91	—42.96
$3Pb^{2+} + 2PO_4^{3-} \rightleftharpoons Pb_3(PO_4)_2$	—7.67	—73.77

spontaneously and require an input of additional energy. Photosynthesis is an example.*

Once the correct way is found for determinating chemical activity by means of ΔZ or ΔF functions, the old Berthelot–Thomson
principle acquires a new meaning. At low temperatures, the
thermal effect of the reaction and its chemical affinity are nearly
the same.

Consequently, the heat of a reaction can be used for an approximate appraisal of the feasibility of the reaction, provided temperatures of the supergene zone are low. Standard heats of reaction,
$\Delta H°$, and standard isobar potentials, $\Delta Z°$, are given in Table 17.
These data were calculated for very definite conditions of the
reaction feasibility for unit active concentrations of ions and 25°C

*An interesting approach to the utilization of isobar potentials was made by Bughelskii
(1963). According to his scheme, reactions have negative values of ΔZ during the
initial stages of sulfide oxidation, i. e., during the formation of sulfates and hydroxides.
At later stages, the isobar potentials of reactions become characteristically positive.
Reactions with negative isobar potentials can start spontaneously in cold climates.
Reactions with positive potentials can take place only in hot climates, and then only in
the presence of compensating minerals. This hypothesis needs verification.

temperature. In general, such conditions differ from natural ones. In nature, HCO_3^- predominates over CO_3^{2-}, HS^-, or S^{2-}, while copper and other metals tend to form hydrated ionic complexes and not the simple ions of the type Cu^{2+} (see p. 44). Unfortunately, there are no precise data available with regard to the mechanisms of mineral precipitation and the corresponding values of $\Delta H°$ and $\Delta Z°$. The data of Table 17 are at best illustrative of the energy analysis method. For the precipitation of sulfides, selenides, sulfates, some cabonates, and some phosphates, both $\Delta H°$ and $\Delta Z°$ have the same sign and change similarly. For the precipitation of carbonates of Mg, Fe, and Zn and the phosphates of Ca, $\Delta Z°$ is negative and $\Delta H°$ positive. Thus, the isobar potential indicates spontaneity for such reactions, even though these reactions are actually endothermal and absorb heat. The Berthelot-Thomson principle does not hold for such reactions. As mentioned earlier, the validity of this principle is relative.

It is possible to determine the equilibrium constant of a reaction from the formula

$$\Delta Z° = -1.364 \log K_p,$$

if its standard isobar potential is known. With this in mind, let us consider again the reaction

$$CaF_2 \rightleftharpoons Ca^{2+} + 2F^-.$$

Its standard isobar potential, $\Delta Z°$, is:

$$\Delta Z°_{reaction} = -\Delta Z° CaF_2 + 2\Delta Z° F^- + \Delta Z° Ca^{2+} = (+277.7) + $$
$$+ 2(-66.08) + (-132.18) = 13.36.$$

Substituting this into the above formula and solving it for K, we have:

$$\log K = + \frac{13.36}{-1.364} = -9.8; \quad K_p = 10^{-9.8}.$$

Since the concentration of CaF_2 in its solid phase is 1, we rewrite:

$$[Ca^{2+}][F^-]^2 = 10^{-9.8}.$$

By means of this formula we have determined the maximum number of F^- ions necessary to accommodate all the calcium ions which are present in the solution Smyshlyaev and Edeleva (1962) have experimentally obtained the value of $(2.7 \pm 0.27) \cdot 10^{-11}$ for the solubility product of fluorite at 25° C. Their figure is somewhat smaller than ours. The difference might be due to various causes, such as the difference between natural and synthetic fluorites, the peculiar modes of occurrence or calcium and fluorine in water, etc. Analogously, knowing the standard isobar potentials of the reaction, we can calculate its equilibrium constants, its solubility products, or its maximal ionic concentrations. It should be kept in mind that such data hold only for 25° C and for the specific mechanism of reaction. Should waters actually contain some other ions, the equilibrium constant and the ionic concentrations would be different. Let us recall that in most cases the ionization power of water and the active concentrations should be reckoned with.

OXIDATION – REDUCTION REACTIONS

In supergene zones, the most important chemical reactions are of the oxidation—reduction type, i. e., those accompanied by a loss or gain of electrons.[*]

The smaller the value of the electronegativity, the easier the atom loses its valence electrons, and, consequently, the stronger is it a reducing agent. The greater the electronegativity, the more pronounced is the oxidizing property of the element. Therefore metals are strong reducers and nonmetals are oxidizers. The most important oxidizing agent is oxygen, which may be atmospheric or hydrospheric. Elements and ions capable of capturing extra electrons include trivalent iron, tetravalent manganese, pentavalent nitrogen, vanadium, arsenic, hexavalent sulfur, selenium, chromium, and molybdenum. These are of lesser importance as oxidizers. Among reducers there should be mentioned bivalent iron, manganese, and sulfur (negative), trivalent chromium and vanadium, and many organic compounds.

The same element may be either oxidizing or reducing depending on the state of its ionization. For example, trivalent iron is an oxidizer, while bivalent iron is a reducer. Tetravalent manganese is an oxidizer and bivalent manganese is a reducer.

Each oxidation reaction is accompanied by a reduction. However, it is customary in geochemistry to study oxidizing and

---

[*]While forming a chemical compound, an atom may lose some electrons and thus become a positive-charged cation, or it may acquire some electrons, becoming a negatively charged anion. In addition, numerous substances are formed without a previous ionization of their atoms. In such cases, valence electrons of the reacting atoms combine into "couples," such couples existing within the spheres of influence of both atomic nuclei. In general, reactions are said to be oxidizing if atoms lose valence electrons and reducing if they gain electrons.

reducing processes separately, investigating only the change in a specific group of substances. For example, one can speak of sulfide oxidation on vein outcrops, although, while copper, sulfur, and other elements are oxidized, the atmospheric oxygen is reduced. The reduction of sulfates dissolved in ground waters is accompanied by the oxidation of organic matter.

Oxidation–reduction ("Redox") reactions are accompanied by either the evolution or absorption of energy. For example, the oxidation of bivalent iron or manganese produces considerable heat:

$$4FeO + O_2 = 2Fe_2O_3 + 131.8 \ cal,$$

$$2MnO + O_2 = 2MnO_2 + 58 \ cal.$$

Under the thermodynamic conditions of supergene zones, these reactions yield energy, which explains the ease of oxidation of bivalent iron and manganese at the earth's surface.

Like all other reactions, those of an oxidation–reduction nature are characterized by isobar potentials, ΔZ. An oxidation–reduction reaction proceeding in an electric battery approximates a reversible reaction. In such a case, the work is equal to the work performed during the chemical reaction, A, which is equal to the negative of the isobar potential:

$$A = \Delta Z .$$

In electrochemistry, the work performed in a galvanic cell is equal to the product of the electromotive force, E, and the current, nF:

$$\Delta Z = nF \cdot E,$$

where n is the ionic charge, and F is the Faraday constant = 96,493.1 coulombs = 23,062 cal/equiv.[*]

Substituting this into the equation for the potential, we have:

$$\Delta Z = RT \ln \frac{a_R'^k \cdot a_S'^l \dots}{a_B'^n \cdot a_C'^m \dots} - RT \ln K$$

and

$$E = -\frac{RT}{nF} \ln \frac{a_R'^k \cdot a_S'^l \dots}{a_B'^n \cdot a_C'^m \dots} + \frac{RT}{nF} \ln K.$$

*All these calculations are for 1 g-atom of an element or for 1 g-ion of a complex ion.

At values of unity for the initial activities, the first term of the equation becomes zero. Such an electromotive force is known as the standard potential, Eo:

$$Eo = \frac{RT}{nF} \ln K = \frac{\Delta Z^\circ}{nF}.$$

Each galvanic cell may be visualized as consisting of two half-cells. Knowing the electromotive force produced by the entire cell, and assuming the potential at the electrode of one half-cell to be zero, the relative potential at the electrode half-cell may be found. In physical chemistry, a hydrogen electrode is used as a standard of zero potential. Its potential is equal to that required to transform one g-molecule of gaseous hydrogen to the ionized state at 25° C and 1 atm:

$$H_2 \rightarrow 2H^+ + 2e.$$

Relative potentials, Eh and Eo, are measured in volts. The formula given above is used for their calculation.

As an example, let us calculate the standard redox potential, Eo, of the oxidation of metallic zinc,

$$Zn \rightleftarrows Zn^{2+} + 2e.$$

The standard isobar potential for this reaction is:

$$\Delta Z^\bullet_{\text{reaction}} = \Delta Z^\circ Zn^{2+} - \Delta Z^\circ Zn.$$

Substituting the isobar potential value found in Latimer's data (Table 15) and noting that the isobar potential of metallic zinc is zero, we get:

$$Z_{\text{reaction}} = -35.184.$$

Substituting this into the formula for Eo, and noting that n = 2 for bivalent zinc:

$$Eo = \frac{\Delta Z^\circ}{nF} = \frac{-35.184}{2 \cdot 23.062} = -0.763 \text{ v.}$$

Therefore, the oxidation of 1 g-atom (65.38 g) of zinc generates a standard potential of −0.763 V at 25° C and 1 atm. Standard redox potentials for a number of metals and ions are (following Latimer):

Strongly acid media

Na $= Na^+ + e - 2.71$	$Sn^{2+} = Sn^{4+} + 2e + 0.15$
Ca $= Ca^{2+} + 2e - 2.87$	$Cu^+ = Cu^{2+} + e + 0.153$
Al $= Al^{3+} + 3e - 1.67$	$U^{4+} \rightarrow UO_2^{2+} + 2e + 0.334$

Strongly acid media (continued)

$Zn = Zn^{2+} + 2e - 0.763$ $\qquad$ $Cu = Cu^{2+} + 2e + 0.337$

$Cr = Cr^{3+} + 3e - 0.74$ $\qquad$ $V^{3+} = V^{4+} + e + 0.40$

$Ga = Ga^{3+} + 3e - 0.53$ $\qquad$ $Cu = Cu^{+} + e + 0.521$

$Fe = Fe^{2+} + 2e - 0.44$ $\qquad$ $2I^{-} = I_2 + 2e + 0.535$

$Cd = Cd^{2+} + 2e - 0.403$ $\qquad$ $Fe^{2+} = Fe^{3+} + e + 0.771$

$In = In^{3+} + 3e - 0.342$ $\qquad$ $Ag = Ag^{+} + e + 0.799$

$Tl = Tl^{+} + e - 0.336$ $\qquad$ $Rh = Rh^{3+} + 3e + 0.8$

$Co = Co^{2+} + 2e - 0.277$ $\qquad$ $Pd = Pd^{2+} + 2e + 0.987$

$Ni = Ni^{2+} + 2e - 0.25$ $\qquad$ $Mn^{2+} \rightarrow MnO_2 + 1.239$

$Sn = Sn^{2+}2e - 0.136$ $\qquad$ $V^{4+} = V^{5+} + e + 1.24$

$Pb = Pb^{2+} + 2e - 0.126$ $\qquad$ $Cr^{3+} \rightarrow CrO_4^{2-} + 1.33$

$H_2 = 2H^{+} + 2e \pm 0.00$ $\qquad$ $Pb^{2+} \rightarrow PbO_2 + 1.45$

$\qquad\qquad\qquad\qquad\qquad$ $Au = Au^{3+} + 3e + 1.50$

$\qquad\qquad\qquad\qquad\qquad$ $Mn^{2+} \rightarrow MnO_4^{2-} + 1.51$

$\qquad\qquad\qquad\qquad\qquad$ $Co^{2+} \rightarrow Co^{3+} + e + 1.82$

Strongly alkaline media

$Ca + 2OH^{-} = Ca\,(OH)_2 + 2e - 3.03$

$Al + 4OH^{-} = H_2AlO_3^{-} + H_2O + 3e - 2.35$

$Th + 4OH^{-} = Th\,(OH)_4 + 4e - 2.48$

$Zn + 2OH^{-} = Zn\,(OH)_2 + 2e - 1.245$

$Cr + 3OH^{-} = Cr\,(OH)_3 + 3e - 1.3$

$Fe + 2OH^{-} = Fe\,(OH)_2 + 2e - 0.877$

$Cd + 2OH^{-} = Cd\,(OH)_2 + 2e - 0.809$

$Co + 2OH^{-} = Co\,(OH)_2 + 2e - 0.73$

$Ni + 2OH^{-} = Ni\,(OH)_2 + 2e - 0.72$

$Fe\,(OH)_2 + (OH)^{-} = Fe\,(OH)_3 + e - 0.56$

$Pd + 2OH^{-} = Pd\,(OH)_2 + 2e + 0.07$

$Mn\,(OH)_2 + 2OH^{-} = MnO_2 + H_2O + 2e + 0.05$

$Cr\,(OH)_3 + 5OH^{-} = CrO_4^{2-} + 4H_2O + 3e + 0.13$

$Co\,(OH)_2 + OH^{-} = Co\,(OH)_3 + e + 0.17$

Alkalies and alkaline earth metals (Na, K, and Ca) possess low standard potentials, since, losing their valence electrons easier than hydrogen, they are transformed into positively charged cations. Being powerful reducers, they decompose water, liberating hydrogen.

Under the conditions of the earth's crust, such elements readily form chemical compounds, and therefore no native sodium, potassium, magnesium, etc., are known. On the other hand, gold, silver, or palladium hold their valence electrons quite firmly and display positive potentials. Such elements occur mostly in the native state. The remaining elements stand between these two groups. Every aqueous solution is characterized by a specific value of the redox potential, i. e., by its ability to oxidize or to reduce different ions. The Eh can be determined with a potentiometer in a manner similar to the determination of pH. The Eh and Eo values can also be calculated for various elements from the isobar potentials of the reaction.

When the concentration of any ion affects the potential of a given solution, such an ion is said to be electroactive. Among the elements and ions controlling the potentials generated in natural water are free oxygen, hydrogen sulfide, bivalent and trivalent iron, bivalent and tetravalent manganese, trivalent and pentavalent vanadium, and organic compounds.

A knowledge of the redox potentials existing in different environments of the supergene zone sheds light on the migration of individual elements, particularly on their concentration or dispersion in subzones of sulfide oxidation, in various ground waters, etc. Once the mode of migration of a single element has been determined, the redox potential of the given solution becomes known, and migration modes of other elements can be determined. For example, low pH values of northern swamps or of some ground waters favor the migration of bivalent iron and manganese. In contrast, oxygenated ground waters of deserts and semideserts are unfavorable to the migration of iron because of its trivalent state.

It follows from the foregoing that oxidation proceeds at lower values of Eh in alkaline media than in acid media. Therefore, oxidation proceeds at a lower Eh in the carbonate rocks of deserts and semideserts than in the acid waters of tundra and taiga swamps. The Eo values correspond to specific conditions of temperature, concentration, and pH. For this reason, the values given in tables do not, as a rule, correspond to actual redox reactions which develop in the supergene zone.

The Eh usually fluctuates between + 0.7 and −0.5 V. Oxygenated surface waters and ground waters containing free oxygen are characterized by a narrower interval: + 0.150 to + 0.700 V. Fissure waters in volcanic rocks display Eh values of + 0.004 to + 0.005 V even at a depth of 250-300 m. The Eh values are considerably below zero, and at times go down to −0.5 V in ground waters of bituminous rocks and oil fields, this being due to microbiologic reduction.

The Eh ranges of oxidation and reduction vary with the elements. For example, at Eh = 0.7 V, a strongly acid medium is reducing for trivalent iron and transforms it into bivalent iron. The same medium is oxidizing for bivalent copper, since its reduction to the monovalent state does not begin until the Eh falls to 0.153 V.

The Eh values at the boundary between oxidizing and reducing media usually vary with pH. Thus, while iron is still bivalent at Eh = 0.5 V in a zone of oxidation of sulfide ore deposits, it would be oxidized to ferric iron at this Eh and a pH of 7−8 within the weathered crust in deserts and semideserts. In other words, the boundary between the oxidation and reduction of iron fluctuates between 0.4 V and 0.6 V in strongly acid media such as zones of sulfide oxidation or the swamps of Siberia. This boundary falls below zero in the alkaline media of deserts.

It follows from the foregoing that certain reducing processes, such as $Fe^{3+} \to Fe^{2+}$, are possible even in the presence of free oxygen dissolved in water, provided that the Eh is positive and varies between 0.4-0.5 V.

Within the supergene zone, oxidation−reduction media can be appraised by means of mineral-indicators. Iron minerals are the most useful in this respect because of their bright coloring. Many minerals of trivalent iron display red, brown, or yellow colors: e.g., hematite, goethite, and limonite. Minerals of bivalent iron are white, blue, or greenish: siderite, vivianite, etc. Such properties of iron, in conjunction with its high abundance in the earth's crust (4.65%), make it a good indicator of oxidizing−reducing conditions in the supergene zone. Red and brown colorations are considered to be due to oxidation, while greenish and bluish-gray tones are due to reduction.

ELECTROCHEMICAL LEACHING OF SULFIDE ORES

In general, metal sulfides are electric conductors. Sulfide ores usually consist of several sulfide minerals, all of which display different conductivities: pyrite, sphalerite, chalcopyrite, etc. When immersed in water, such minerals form microgalvanic couples. According to Sveshnikov, the electromotive force generated by such couples may reach 0.3-0.4 V. At the same time, metal sulfides serving as anodes dissolve. For example, if an ore consists of pyrite, galena, sphalerite, and chalcopyrite, lead and zinc will go into solution, while copper and iron will not be corroded to any appreciable extent. In such a case, pyrite and chalcopyrite become cathodes while galena and sphalerite become anodes.

Experimentally, S. Okhasi (1953) has shown that distilled water is enriched in lead upon immersion of the following couples: galena-sphalerite, galena-chalcopyrite, and galena-pyrite.

Such phenomena can develop in any medium—alkaline, neutral, or acid—the presence of oxygen not being necessary. Thus, the aqueous haloes of lead and zinc may develop in deep-seated conditions, provided that metal sulfides are present.

Rasskazov (1963) has reported electrochemical leaching of sulfide-arsenide mixtures, where cobalt and nickel go into solution while iron and lead remain as residual sulfide and arsenides. He noted the following reactions:

On the anode: $As_2 - 4e = Co^{2+}, Ni^{2+}, As°$;
On the cathode: $MeS + 2e = Me° + 2S°$.

Electrochemical leaching also occurs in the oxidation zones of sulfide ore deposits. There, minerals possessing the most negative electrode potentials dissolve most extensively. For example, galena dissolves more energetically when mixed with pyrite, while the pyrite is hardly affected at all.*

STABILITY RANGES OF MINERALS AND CHEMICAL COMPOUNDS

The properties of minerals and chemical compounds may be graphically portrayed as functions of their various parameters: Eh, pH, CO_2 gas pressure, etc. The Eh–pH relationships are used most commonly, since these properties are the most important characteristics of natural waters.

The Eh–pH diagrams represent ranges of stability of minerals or compounds, and are calculated from thermodynamic data for conditions of equilibrium. However, many chemical reactions develop very slowly in the supergene zone, particularly if the climate is cold. The thermodynamical equilibrium may not ensue quickly enough. Hence, some discrepancies are to be expected between actual observations and theoretical deductions. For example, neither pyrite nor any other sulfide is in equilibrium with oxygenated water; therefore, they should oxidize. In polar regions,

---

*According to Udodov et al. (1962), oxidation advances faster than leaching. He cited a sample of ground water from a 200-m depth in a drillhole at a Bukhtarinskii copper-cobalt ore deposit. On analysis, it yielded only a trace of copper but $7.2 \cdot 10^{-6}$ g/liter of Zn and $2.3 \cdot 10^{-6}$ g/liter of Pb, the pH being 7.0.

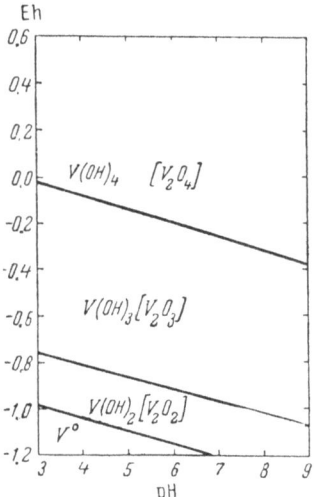

Fig. 12. Range of stability of vanadium hydroxides
(after Garrels, 1962).

however, the oxidation reaction is very slow, and sulfide outcrops
escape oxidation. In addition, while feldspars are unstable thermo-
dynamically at the earth's surface, the speed of their weathering
is very low in arid climates; therefore, arkosic sands remain
unaltered for a long time. These instances limit the usefulness of
Eh—pH diagrams to an illustrative value only. Even at that, their
value is considerable since they portray changes in parameters such
as temperature or pressure.

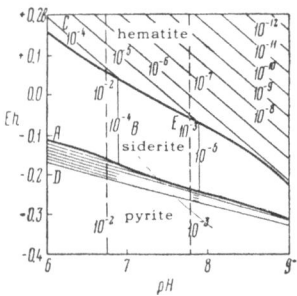

Fig. 13. Stability diagrams for hematite, siderite, and
pyrite with superposed activities of total iron (ferrous +
ferric) in thin lines and activities of calcium ion in dashed
lines (after Crumbein and Garrels, 1960).

Stability ranges of vanadium compounds are shown in Fig. 12 for 25° C and 1 atm. It is clear that neither metallic vanadium nor its bivalent ion can possibly be in equilibrium in supergene environments. In fact, neither one can possibly form, since they decompose water. Minerals of trivalent vanadium are stable only under reducing conditions and those of pentavalent vanadium only oxidizing conditions. At lower Eh, compounds of pentavalent vanadium are more stable in alkaline media than in strongly acid ones.

Stability ranges of iron minerals, calculated for sea water, are shown in Fig. 13. With certain approximations, such data may be applied to other supergene environments. For instance, pyrite forms at negative Eh values and more readily in reducing environments of alkaline media than in acid media. Hematite remains stable in the alkaline environments of deserts even at Eh values below zero. On the other hand, it is stable at pH 6 only in definitely oxidizing environments, transforming into siderite whenever the Eh is lowered to +0.1 V.

Comparing Figs. 13 and 14, we can see that manganese oxides exist in almost the same circumstances as hematite, while rhodochrosite, $MnCO_3$, is stable only in conditions more reducing than permit the existence of siderite; therefore, it can occur in association with pyrite.

Stability ranges of ions and hydroxides of elements in groups IVa, Va, VIa, and VIIa of the periodic table are given in Fig. 15. These data indicate that compounds of hexavalent chromium cannot exist in strongly acid media because of the absence of a sufficiently high potential (in excess of 1.2 V) in the supergene zone. Their

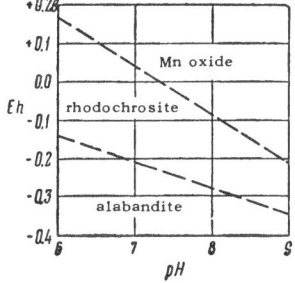

Fig. 14. Approximate ranges of stability of manganese oxides, rhodochrosite, and alabandite (after Crumbein and Garrels, 1960).

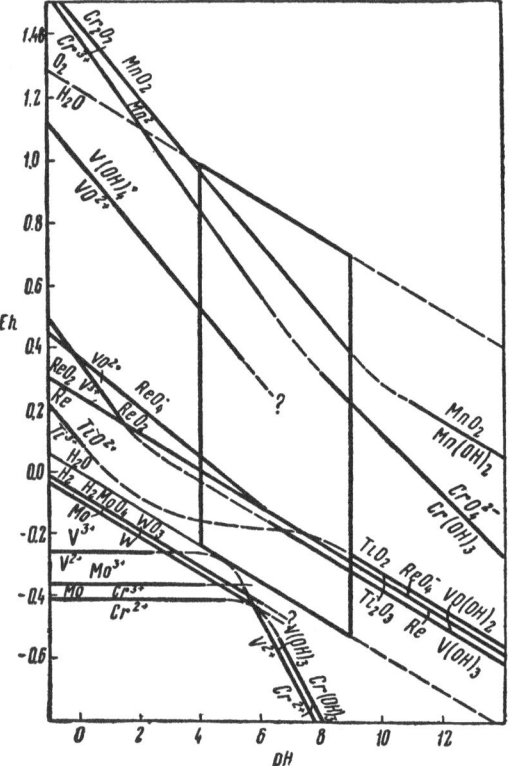

Fig. 15. Stability ranges of ions and hydroxides of rare metals of groups IVa, Va, VIIa of the periodic system at 25° C, 1 atm, and a cationic concentration of 10^{-7} mole/liter (after Krauskopf, 1958).

existence would be more probable in alkaline environments, since the transformation of Cr^3 to Cr^6 is accomplished in nature at lower Eh values—e. g., 0.4 V at pH 8. This conclusion agrees with observations of tarapakaite, K_2CrO_4, and other soluble dichromates of Chilean deserts.

Stability ranges of ions and hydroxides of groups VIII and Ib of the periodic table are given in Fig. 16. It follows that while native copper can exist in the supergene zone, native cobalt and nickel cannot because the needed potentials are too low to be found in the supergene zone. A comparison of graphs shows that the reduction of bivalent to monovalent copper takes place in a more reducing environment than does the reduction of trivalent iron or tetravalent manganese or vanadium. In other words, reduced compounds of these elements may occur in association with oxidized compounds of copper. This explains the contemporaneous migration of bivalent iron, manganese, and copper in acid waters of tundra and taiga.

The partial pressure of CO_2, H_2S, or some other gas has been used in geochemistry as a third parameter. Stability ranges of uranium compounds are shown in Fig. 17 in terms of Eh–pH–CO_2 coordinates.

LECHATELIER PRINCIPLE

The LeChatelier–Brown thermodynamical principle can be used to determine the direction of a given reaction in the supergene zone.

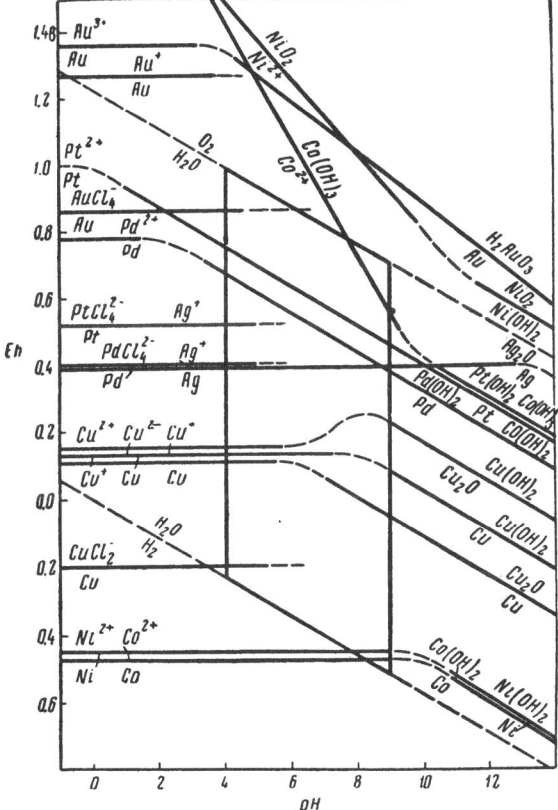

Fig. 16. Stability ranges of ions and hydroxides of rare metals of groups VIII and 1b of the periodic system at 25°C, 1 atm, and an ionic concentration of 10^{-7} mole/liter (after Krauskopf, 1958).

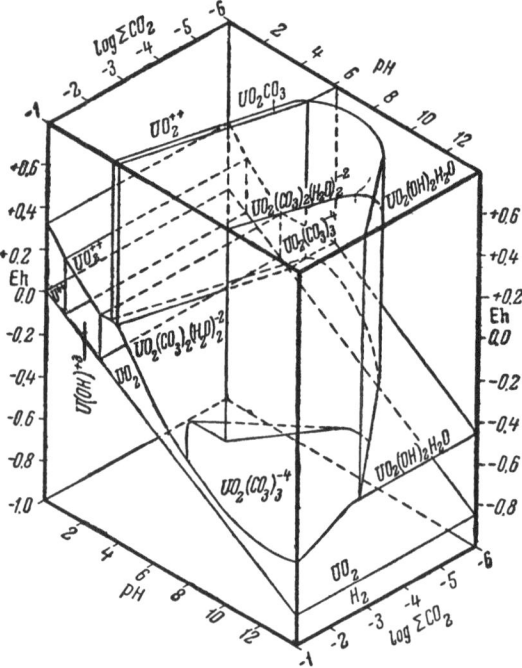

Fig. 17. Stability ranges of uranium compounds, calculated as functions of Eh, pH, and the total dissolved carbonates at 25° C, 1 atm of total pressure (after Garrels, 1962).

This principle is defined in the Handbook of Physics and Chemistry as follows:

> "If some stress is brought to bear upon a system in equilibrium, such a change occurs that the equilibrium is displaced in the direction which tends to undo the effect of the stress."

It should be emphasized that the LeChatelier principle is applicable only to systems in true equilibrium. A salt lake with salts precipitated at its bottom is an example of such a system. As temperatures fall in autumn and early winter, the first salts to precipitate are those which crystallize with the greatest evolution of heat. It would appear as if the system tried to offset the coming change in the external conditions. Conversely, when temperatures rise in spring and early summer, the first salts to dissolve are those which absorb the most heat while dissolving. The LeChatelier principle cannot be utilized to determine the direction of the process in a system not in equilibrium — such as primary silicates — natural supergene waters, or sulfides – atmospheric oxygen. Despite the prevailing opinion, the significance of the LeChatelier principle in analyses of supergene processes is rather small.

Pallman (1933) tried to explain the general direction of the weathering process on the basis of this principle. His line of reasoning was as follows: Primary minerals formed at elevated temperatures and pressures and then placed at the earth's surface at low temperatures and pressures should, according to the LeChatelier principle, alter in the direction opposite to any change in existing external conditions. Therefore, weathering reactions should produce compounds with a greater molecular volume and with the evolution of heat to offset the decreases in pressure and temperature. Pallman's view was publicized greatly Rode (1937). A similar view was taken by Pustovalov (1940) in his well-known treatise:

> "When investigating sedimentation phenomena, we meet with manifestations of this important law [the LeChatelier principle] at practically every step. Actually, magmatic minerals, formed at high pressures and then exposed to a pressure of only 1 atm at the earth's surface, should evolve into products with a much greater total volume than the original. A lowering of the temperature, which is experienced by magmatic minerals on their exposure at the earth's surface, is compensated for by the heat evolved during their decomposition."

However, no weathering environment represents a system in equilibrium. Supergene natural waters, soil, or weathered crust are not the systems in which feldspars, olivines, and other primary minerals formed. Therefore, it is erroneous to talk about applying the LeChatelier principle to these cases.

Lebedev (1957) justifiably warns against any simplified interpretation of the LeChatelier principle, and cites numerous examples of errors incurred as result of such an interpretation.

FERSMAN'S GEOENERGETIC THEORY

A. E. Fersman sketched a new method for analyzing the energy involved in natural processes when he postulated his geoenergetic theory. Noting that in the earth's crust many processes run in the direction of diminishing ΔZ and ΔF, he concluded that magmatic minerals crystallize from melts in the order of diminishing lattice energies, such energy being characteristic of natural processes.

Furthermore, Fersman assumed that lattice energy is an additive value which consists of energies due to each participating ion. He named this energy the energy constant of the ion, EK. This constant is as typical of an ion as is, for example, its polarization.

By a series of further assumptions, Fersman calculated values of EK for most ions and thus adapted his theory to practice. Here, we shall express EK in arbitrary units which can be converted into kg-calories if multiplied by a factor of 256.1.

To calculate EK values of cations and anions, Fersman derived the following formulas:

$$EK_{cation} = \frac{W^2}{2R} \, 0.75 \, (R + 0.20),$$

$$EK_{anion} = \frac{W^2}{2R},$$

where W is the valence and R the radius of the ion.

Another constant, VEK, is obtained by dividing the valence into the corresponding EK.

The EK concept was further developed by Gurevich and Pavlov (1963) who introduced as a new concept the total EK of natural water calculated for one liter and proportional to the total mineralization. They also proposed the use of the average specific EK calculated per one g-equivalent of dissolved salts.

By means of his geoenergetic theory, Fersman explained the order of consecutive crystallization of different elements from cooling magmas all the way down to the postmagmatic stage. Fersman utilized the concepts of lattice energy, EK, and VEK to evaluate some supergene processes.

Sequence of the accumulation of elements in water basins

Cation	Na	K	Rb	Cs
EK	0.45	0.36	0.33	0.30
Anion	Cl	Br	I	
EK	0.25	0.22	0.18	

The greater the energy of the crystal lattice of a mineral and, consequently, the greater its EK, the less soluble it is. Therefore, elements with high EK values and small ionic radii — such as Si, Al, and Fe — accumulate in the form of residual weathering products. Elements with small EK values and larger ionic radii are easily leached from soil or eluvium and go into solution. EK values of different ions are given in Table 18.

We note, for example, that the PO_4^{3-} ion adds 384 cal on its incorporation into a crystal lattice, while the CO_3^{2-} ion adds only 200 cal and the Cl^- ion 64 cal (1.50, 0.78, and 0.25 arbitrary EK units, respectively).

In general, solubility and mobility decrease from chlorides through carbonates to phosphates. Basic halides are characterized by lattice energies of only 180 to 140 cal, the lowest known in the supergene zone.

There are numerous phenomena which cannot be explained on the basis of this theory. According to Polynov (1934), for example, cations extracted during the weathering of primary minerals follow the sequence: Ca and Na > Mg > K; whereas the sequence in keeping with the geoenergetic theory would be: K > Na > Ca > Mg, as the data of Table 18 indicate. The discrepancies, however, may be due to the effect of the biologic factor. Another item not accounted for by the geoenergetic theory is the concentration of substances in solutions, which exerts a very definite influence on the aqueous migration of elements.

Table 18. EK Values for Some Ions Typical of the Supergene Zone (after Fersman, 1937)

Anion	EK	EK × 256.1	Cation	EK	EK × 256.1
OH^-	0.37	95	H^+	1.1	282
CO_3^{2-}	0.78	200	Na^+	0.45	115
NO_3^-	0.18	49	Mg^{2+}	2.10	538
PO_4^{3-}	1.50	384	Al^{3+}	4.25	1265
SO_4^{2-}	0.70	180	K^+	0.36	12
Cl^-	0.25	64	Ca^{2+}	1.75	448
CrO_4^{2-}	0.75	172	V^{3+}	5.32	1362
F^-	0.37	95	Cr^{3+}	4.75	1216
Br^-	0.22	56	Mn^{2+}	2.00	512
I^-	0.18	46	Fe^{2+}	2.12	543
			Fe^{3+}	5.15	1320
			Co^{2+}	2.15	553
			Ni^{2+}	2.18	558
			Cu^{2+}	2.10	538
			Zn^{2+}	2.20	563
			Ag^+	0.60	154

We can summarize the foregoing by saying that in the supergene zone, the degree of migration of the elements is influenced by the EK and the VEK in part but not exclusively. Insofar as the biogenic migration of chemical elements is concerned, the greatest role is played by the nonionic bonds, and the concepts of EK and VEK are superfluous.

In conclusion, much caution should be exercized when applying the geoenergetic theory to studies of aqueous migration.

PRINCIPLE OF RETARDATION OF REACTIONS

In natural systems, the exact, stoichiometric proportions of reagents are rare, usually being an excess or a deficiency of some reagents. Such an unbalance is largely due to different abundances and different solubilities of these substances. Therefore, the problem of reagent concentration is of utmost importance in geochemistry.*

As specified by the law of mass action, any supergene process may yield different results depending on the initial concentrations of the reacting substances. Two general cases are:

1. The given reagent is in excess in relation to the amounts of other reagents, there being more than enough of it for any possible reaction.

2. The given reagent is in insufficient amount.

As an example, there is an excess of free oxygen in the upper part of the oxidation zone of sulfide ore deposits. All primary ore minerals should oxidize completely. In the lower levels of the same zone, oxidation is feasible only at the expense of the oxygen dissolved in ground water. Its oxygen is insufficient for a simultaneous oxidation of all sulfides.

Oxygen deficiency is characteristic of a number of supergene environments, particularly in the subzone of catagenesis with its water-bearing strata. A deficiency of hydrogen sulfide might be encountered in such places, the amount of this gas being insufficient to precipitate all the available metals.

---

*In contrast, laboratory chemists start their analyses with carefully calculated quantities which are then increased by small amounts to allow for the completion of the reaction in accordance with the law of mass action.

In view of the ionic composition of natural waters, we can further distinguish two classes of ionic deficiency: a deficiency of (1) anions and (2) cations. The first of these is particularly common. For example, the content of PO_4^{3-}, VO_4^{3-}, AsO_4^{3-}, F^-, and other anions is usually smaller than the corresponding number of cations available for the formation of insoluble compounds (such as Ca_3PO_4). Cationic deficiency is encountered occasionally in certain environments.

In general, a deficiency of elements is a typical feature of the supergene zone. Because of such a deficiency, not all theoretically possible reactions can actually take place. For instance, not all the sulfide ore minerals will be oxidized by the oxygen dissolved in ground water.

The question of which of the possible reactions would be the probable one is answered on the basis of chemical affinity. Let us recall that chemical affinity is measured by the work, A, performed during a reaction, and that $A = -\Delta Z$.

Should one of the necessary reagents be deficient in a given supergene environment, only those reactions can take place which are characterized by the greatest amounts of work performed, or, which is the same, by the lowest isobar potentials. Such reactions will actually prevent the realization of reactions with larger potentials simply by using up the available reagents. This premise was formulated by the author in 1941 and is referred to as the principle of retardation of chemical reactions (Perel'man, 1947, 1961b).

Since in the conditions of the supergene zone the heat of reaction is close in value to the isobar potential, the above rule can be simplified to fit many cases in the supergene zone: When some reagents are deficient, those reactions will actually take place which are accompanied by the greatest evolution of heat. Because the isobar potential is frequently proportional to the solubility product, it can be further stated that reaction will actually take place which is characterized by the lowest solubility product.

Deficient as the reagents may be, they are continually replenished in the supergene zone. In geologically long periods of time, their total amounts may become considerable. However, at any single moment their amounts are small enough to effect the retardation.

Reagent deficiency is rare in laboratory practice. Sometimes it is induced artificially as in fractional precipitation. For example, when $AgNO_3$ (the deficient reagent) is added drop by drop to a solution containing equal concentrations of Cl^- and I^-, there first precipitates pure AgI because at 25° C

	Isobar potential	Solubility product
AgI	-21.93	$1 \cdot 10^{-16}$
AgCl	-13.30	$1 \cdot 10^{-10}$

Thus, the reaction of $AgCl$ precipitation is retarded by the reaction of AgI precipitation. As the iodide, I^-, anions are being used up, the concentration of Ag^+ cations will keep increasing, since the solubility product of AgI is a constant. At the moment when equal amounts of silver ions would be needed to form both AgI and $AgCl$, precipitation of both salts should start simultaneously.

In natural surroundings, the depletion of ions may not be balanced immediately by their supply, e. g., new portions of ground water flow may not be arriving continually. Therefore, the dynamical nature of the supergene zone makes it more susceptible to retardation processes. Reagent deficiency has a definite effect on many geochemical processes, making them differ from analogous reactions in the laboratory. In fact, the paragenesis of elements is one of the peculiarities of supergene ores produced by retardation (see Chapter 15).

One of the problems frequently confronting the investigators may be stated in the following manner: A water flow containing variable amounts of different chemical elements is approaching an area where the precipitant is deficient: what will the sequence of precipitation of those elements be? It is assumed that the concentration of the precipitant is to be prorated against all participating elements. Let us use an idealized example to illustrate the solution of this problem.

We assume that the precipitation environment is characterized by water having an active concentration of S^{2-} ions $a_{S^{2-}}$, while the water flow carries Cu^{2+}, Pb^{2+}, and Zn^{2+}. Solid CuS precipitates as a result of a mingling of these two types of water. What will the value of the isobar potential of this reaction be?

The reaction is given by:

$$CuS \rightleftarrows Cu^{2+} + S^{2-},$$

then

$$\Delta Z_{reaction} = RT \ln \frac{a_{Cu^{2+}} \cdot a_{S^{2-}}}{a_{CuS}} - RT \ln \left(\frac{a_{Cu^{2+}} \cdot a_{S^{2-}}}{a_{CuS}} \right) \text{equilibrium} \cdot$$

Since the concentration of CuS in the solid phase is 1, we can re-write:

$$\Delta Z_{reaction} = RT \ln \frac{a_{Cu^{2+}} \cdot a_{S^{2-}}}{1} - RT \ln SP = RT \ln a_{Cu^{2+}} +$$

$$+ RT \ln a_{S^{2-}} - RT \ln SP \quad CuS.$$

Similarly we have

$$\Delta Z_{reaction} = RT \ln a_{Pb^{2+}} + RT \ln a_{S^{2-}} - RT \ln SP \quad PbS.$$

The second term on the right, $RT \ln a_{S^{2-}}$, is the same for both equations, and, therefore, concentrations of S^{2-} are the same in both cases. For a given precipitation environment, actual differences between isobar potentials depend on the value of the solubility product SP of the precipitated mineral and the active concentration of the element–precipitant.

Should the solubility of the precipitated mineral be small, its solubility product would also be small. It would be smaller than $a_{Cu^{2+}}$, $a_{Pb^{2+}}$, etc. In such cases, the isobar potential would largely depend on the solubility product.

Assuming the concentration of S^{2-} ion to be 10^{-10} mole/liter and the concentrations of copper and zinc in the water flow to be at 10^{-6} mole/liter, the isobar potentials of precipitation reactions will be:

Reaction	ΔZ	$\Delta Z°$	SP
$Cu^{2+} + S^{2-} \rightleftarrows CuS$	−27.50	−49.33	$10^{-36.2}$
$Zn^{2+} + S^{2-} \rightleftarrows ZnS$	−12.49	−34.32	$10^{-25.2}$

The values are obviously lower for CuS than for ZnS.

Chapter 5

Migration of Colloids

The migration of colloids is very common in the supergene zone and constitutes a large percentage of atomic migration. In this chapter, we will consider four topics which are of utmost importance to the geochemistry of epigenesis: (1) the occurrence of colloids in the supergene zone, (2) the migration of substances in the colloidal state, (3) sorption, and (4) the principal groups of colloidal minerals.

OCCURRENCE OF COLLOIDS IN THE SUPERGENE ZONE

Oxygen, silicon, aluminum, calcium, sodium, potassium, magnesium, titanium, sulfur, manganese, and chlorine constitute 99% of the solid earth crust. Six of these — sodium, calcium, potassium, magnesium, chlorine, and sulfur—constitute 11% of the lithosphere. During weathering, these six elements are removed with relative ease from the lattices of primary minerals and then form more or less soluble salts (such as Na_2SO_4). A portion of their atoms is carried in natural waters in the form of simple or complex ions, and another portion is sorbed by organisms and colloids. Compounds of silicon, aluminum, iron, titanium, and manganese, which constitute about 84% of the lithosphere, are soluble to a lesser extent, and consequently their contents in natural waters or in organisms are less. During weathering, these elements form hydrated colloidal gels which eventually dehydrate and recrystallize.*

Inasmuch as elements of this second group are more abundant in the earth's crust than are elements of the first group, colloids

---

*Should the recrystallized structure be discernible to the naked eye or under a microscope, such a substance is known as a "metacolloid."

and metacolloids predominate over simple salts in the solid products of weathering. Some weathered crusts, soils, or continental sediments contain no simple salts at all. Practically the entire mass of the solid phase of a soil or a weathered crust may be in a colloidal state or may have passed through a colloidal state. Numerous clay minerals, hydroxides of aluminum, iron, manganese, and silicon, humates, etc., are examples.

It has been shown by Dobrovol'skii (1964) that in the supergene zone colloidal minerals can metasomatically replace feldspars and other clastics. This so-called "supergene metasomatism" is especially typical of humid regions where colloidal hydroxides of iron or manganese replace clay minerals, clastic silicates, aluminosilicates, and even quartz.

Metasomatism is less widespread in arid regions where calcite replaces clay minerals while the alteration of clastics is but slight. During supergene metasomatism, colloids sorb some replaced materials, while others are carried away in surface or ground waters. Enormous quantities of elements are thus moved in the supergene environments of humid regions.

The use of electron microscopy and X-ray and other modern techniques has permitted systematic study to be made of the heretofore confused field of "mineral colloids." A new branch of mineralogy, "colloidal mineralogy," is being developed and promises to be of considerable practical value (Chukhrov, 1955).

MIGRATION OF SUBSTANCES IN THE COLLOIDAL STATE

A large number of chemical elements can migrate in the form of colloids, forming sols (colloidal solutions). Colloidal migration is particularly common in regions with humid climates and acid waters enriched in organic substances.

A study of melt waters in permafrost regions of northwestern Siberia indicates that most of the dissolved compounds of Mn, As, Zr, Mo, Ti, V, Cr, and Th migrate in colloidal form. According to Shvartsev (1963), Cu, Pb, Zn, Ni, and Co may migrate colloidally. According to Udodov (1962), the migration of Zr, Sn, and Ti in Siberia is essentially colloidal.

Having acquired mobility, colloidal minerals migrate in soils and weathered crusts of humid regions in the form of dispersions. Their particles become oriented during precipitation and develop "metakinetic" microstructures—thin films of clays coating the walls of cavities, fissures, etc. (Dobrovol'skii, 1964). V. Kubieva, a German topologist, has named such films "plasmas." Their geochemical significance is not clear, but they may have the role of sorption barriers which retard the mobility of some elements and thus affect the composition of water.

In arid climates, there is much calcite in soils and continental sediments. Their ground waters have a weakly alkaline reaction, and, as a rule, do not contain any organic acids. All this aids the colloidal migration.

SORPTION

Colloids are capable of sorbing (assimilating) ions and molecules which are present in natural waters in concentrations not exceeding their solubility products. Sorption is a mechanism of precipitation from undersaturated solutions. Elements like Li, Rb, and Tl do not form saturated solutions in the supergene zone, and sorption by colloids or living matter is the means by which they go from solution into the solid phase.

It is customary to distinguish between adsorption and absorption. The first means sorption assimilation on the surface of colloids; the second, assimilation by the entire mass of a colloid.

The more important sorbents include humates (which are found in soils, oozes, and rocks), clay minerals, various forms of silica, and hydroxides of iron, aluminum, and manganese. By sorbing dissolved metals, such substances become enriched in copper, nickel, cobalt, barium, zinc, lead, uranium, thallium, etc.

It was established by Verigina (1962) that the silty fraction of podzol soils at Klinsko-Dmitrovsk is enriched in copper, zinc, and cobalt, and that the silty fraction of the underlying clay carries 60 to 80% of the total microelements contained in the rock.

Dobritskaya et al. (1962) showed that copper, zinc, cobalt, and molybdenum increase in soils with the increase in clay fraction. Zhuravleva (1962) noted similar contents of Cu and Zn in alluvial soils of the Chita state.

The following relationships were given by Shvedas (1963) for Lithuanian soils:

where x is the copper content of the soil in mg/kg; y is the amount of physical clay, %; Z is the sum of sorbed bases, equiv/100 g; and a and b are constants.

The adsorption of cations and anions from natural waters is known as polar (ionic) sorption. Negatively charged colloids sorb cations and positively charged colloids sorb anions. These reactions bear the character of exchange since the equivalent quantity of cations will be taken out of the solid phase and put into the solution to replace the sorbed ones. In nature, the commonest colloids are negatively charged, and sorb cations. As a mechanism of emplacement, sorption is important in the formation of haloes about ore deposits. All negative colloids contain a certain number of exchange cations which condition the exchange capacity of a colloid. This capacity is measured either in weight-% of a dry colloid or in milliequivalents per 100 g. The quantity of sorbed cations rarely exceeds 1%.

Thus, there are exchangeable cations in clays or in any rocks containing colloids. The nonexchangeable cations are situated at the vertices of crystal lattices and can go into solution only after the destruction of such lattices. Their mobility is seldom great, since this mode is relatively inert. On the other hand, exchangeable cations easily move from solid rocks into solution, this being a mobile mode of occurrence.

Adsorption is governed by the law of mass action. It increases in proportion to the concentration of cations in water. Polyvalent cations are sorbed the most energetically, the energy of sorption being $R^{3+} > R^{2+} > R^+$. For ions of the same valence, the energy of sorption increases with the atomic weight and ionic radius: $Li < Na < K < Rb < Cs$. Therefore, potassium is more energetically sorbed by colloids than sodium and is more firmly retained.

The large uranyl cation UO_2^{2+} is easily sorbed by lignite, phosphorites, kaolin, and montmorillonite. Some authors believe that the nonmineral form of uranium occurring in clays and coals has been sorbed.* The energy of sorption of bivalent ions by montmorillonite decreases in the following manner: $Pb > Cu \geq Ca > Ba \geq Mg > Hg$; sorption by kaolin follows the order: $Hg > Cu > Pb$. The reverse displacement follows the order: $Mg \geq Ba > Ca \geq Cu >$

---

*According to Salai (1959), uranium is adsorbed by humic acid from very dilute solutions in the pH interval from 3 to 7. Thus, as much as 0.01% uranium can accumulate in peat although the solution contained only $1 \cdot 10^{-5}$ g/liter. As much as 10% uranium may be found in a variety of peat known as dopplerite.

Pb for montmorillonite, and Pb > Cu > Hg for kaolin (as reported by Antipov-Karataev and Kader, 1947).

Positively charged colloids, such as hydroxides of aluminum, titanium, chromium, and, in part, iron, are less common in the supergene zone. Such anions as PO_4^{3-}, VO_4^{3-}, AsO_4^{3-}, and SO_4^{2-} may be sorbed from natural waters by hydroxides of iron or aluminum. Such sorptions account for the presence of P, V, or As impurities in brown iron ores. Exchange adsorption of SO_4^{2-} and Cl^- has been established for laterites and red soils.

Exchange adsorption is reversible. It is possible that previously adsorbed ions will be dropped by colloids as the composition of natural waters changes. This phenomenon is known as desorption. However, if ions form a stable compound on having been sorbed by the surface of the colloid, such adsorption, known as chemisorption, is not reversible.

As a result of chemisorption and subsequent crystallization, there develop in brown ironstones such minerals as fervanite, ferromolybdite, various iron phosphates, etc.

"Nonpolar adsorption" involves the sorption of entire molecules. The adsorption of gases and vapors by soils and clays is an example.

Krauskopf (1963) has investigated the adsorption of metals from sea water within the pH range 7.7-8.2 at 18°-23°C. Initial concentrations of metals were on the order of $n \cdot 10^{-4}$ g/liter. Clays, peat, apatite, and hydroxides of iron and manganese were tested as adsorbents. It was established that Zn, Cu, Pb, and Co could be adsorbed to the extent of 90%, the corresponding equilibria being established within a few days. Much depended upon the nature of the adsorbent — while 90% of available nickel was sorbed by ferric hydroxide, apatite extracted only 50 to 60%.

Hydrated manganese dioxide, $MnO_2 \cdot nH_2O$, was found to be the most efficient adsorbent; it was capable of sorbing both cations (Ni, Co, etc.) and anions (W, Cr). Pentavalent vanadium was sorbed most energetically (by ferric hydroxide) and silver most sluggishly (by organic matter).

It was concluded that the principal role in the extraction of metals from sea water was played by sorption, particularly for Cu, Zn, and Pb.

Sorption is an important mechanism of formation of secondary dispersion haloes. Haloes of ore deposits in the eastern Baikal region contain lead, zinc, and molybdenum co-precipitated with ferric or manganese hydroxides (Krischuck, 1963). According to

Polikarpochkin (1962), ferric hydroxides or organic matter sorb lead in amounts up to 1%. In general, the capacity of different substances to sorb lead and zinc decreases in the order: lignite, phosphorite, ferric hydroxides, kaolin (Rozhkova et al., 1962).

Dispersion haloes are formed by metal-bearing solutions filtering through rocks which contain colloids. In resulting exchange reactions, ore elements are sorbed by colloids of the rock, which at the same time release some of their exchange cations into the solution. Thus, ore elements concentrate in a halo without forming any specific minerals. As an example, the following exchange reaction can be expected between a copper-bearing solution and clay, G, containing exchange cations of calcium:

$$\frac{G\text{-}Ca^{2+}}{\text{solid}} + \frac{Cu^{2+} + SO_4^{2-}}{\text{dil. solution}} \rightleftarrows \frac{G\text{-}Cu^{2+}}{\text{solid}} + \frac{Ca^{2+} + SO_4^{2-}}{\text{dil. solution}}.$$

The copper cation displaces the exchange calcium out of clay, and, as a result, the solution begins to contain less copper and more calcium. Since this reaction is reversible, no completion of the replacement can be expected during any short time interval. During long periods of time, clays will eventually accumulate copper in amounts of a fraction of 1%. No insoluble salts of copper are precipitated, nor are waters saturated with copper.

The incorporation of ore elements within the dispersion haloes is controlled by the character of the secondary minerals composing the rock in which the halo is developing. If the secondary minerals are kaolin, the chances of metal accumulation are slight, and the ore elements will keep migrating. If the secondary minerals are bentonitic, the ore elements will be sorbed in the form of a halo.

PRINCIPAL GROUPS OF COLLOIDAL MINERALS

Humus is the name given to black colloidal substances of complex organic composition and acid nature. Forming during the decomposition of dead vegetation, it constitutes as much as 20% of soils and is an important component of lignite and peat. Its content in sedimentary rocks is highly variable — even the light-colored loesses contain up to a few tens of one percent. Surface and ground waters carry humic substances both in solution and in suspension.

Its adsorptive capacity is great, measurable in hundreds of

milliequivalents per 100 g. Since it is negatively charged, it adsorbs cations, and soil humus usually contains sorbed calcium, magnesium, and, occasionally, hydrogen or aluminum. Forest humus generally runs higher in copper, nickel, cobalt, zinc, silver, beryllium, and other elements which accumulate biogenically. If developed near ore deposits, humic substances accumulate ore elements. An admixture of ore elements is usually found in mineral coal: molybdenum, uranium, nickel, cobalt, lead, titanium, scandium, etc. Such occurrences may be of commercial value (e. g., uraniferous coals).

Humic substances are mobile in alkaline media. They easily migrate in soda waters which drain off alkali flats or weathering petroliferous strata. They are also soluble in waters of acidic swamps, to which they impart a characteristic dark brown color. Humus loses its mobility in neutral and weak alkaline media, but may still migrate in the form of mechanical suspensions.

One of the various mineral colloids, negatively charged silica gel, sorbs many metals, including copper and cobalt. It occurs in many soils, as well as in sedimentary rocks and in the oxidation zones of ore deposits. Opals, $SiO_2 \cdot nH_2O$, are typical gels of silica, while chalcedony and some quartz are metacolloids.

Colloidal and metacolloidal hydroxides of manganese are also negatively charged. They occur in the form of black films, smudges, dendrites, and earthy masses in soils, lake and bog sediments, and weathered crusts. They are common in old water channels and constitute the bulk of sedimentary manganese ores. Mineralogically, they are mixtures of manganite, $Mn_2O_3 \cdot H_2O$, pyrolusite, MnO_2, vernadite, $MnO_2 \cdot H_2O$, and wad, a colloidal mixture. Manganese hydroxides and especially wads readily extract from solution such elements as lithium, copper, nickel, cobalt, zinc, radium, barium, tungsten, silver, gold, and thallium. Not infrequently, they contain some silica and alumina, up to 2% BaO, and up to 12% alkalies.

Silicates comprise a large group of colloidal and metacolloidal minerals and include clay minerals. All carry negative charges and are capable of absorbing cations. Although the particle size of clay minerals is generally larger than that of true colloids, they possess many properties which are characteristic of colloids — for example, the exchange capacity.

Clay minerals can be divided into several groups. Minerals of the montmorillonite group form mostly in arid climates. They swell in water and possess considerable sorptive capacities: 50 to

150 milliequivalents per 100 g. To this group belong montmorillonite, beidellite, nontronite, volchonskoite, etc. Montmorillonites
occurring in the vicinity of ore deposits may be zinc bearing
(sauconites) or copper bearing (medmontites).

Minerals of the kaolinite–halloysite group form under the conditions of acid weathering in hot and humid climates. Their
sorptive capacity is negligible and their swelling in water is
insignificant.

Hydromicas, widely distributed in soils, weathered crusts, or
continental sediments, hardly swell at all in water, but display
considerable sorptive capacities: 20 to 40 milliequivalents per
100 g.

Metacolloidal silicates include the palygorskite group, which
is rich in magnesium cations and which forms in alkaline media.

A special group consists of minerals with mixed-layer lattices.
The commonest minerals of this type are combinations of alternate
layers of hydromica (illite) and montmorillonite or if chlorite and
vermiculite (Parfenova and Yarilova, 1962). According to Dobrovol'skii (1964), the mineral composition of the "Minus 1 mm"
fraction of quaternary deposits of the USSR consists of illite intercalated with layers of montmorillonite.

The chemical composition of silicates is never constant but
varies depending on the amount of water, exchange ions, etc. A
general idea of their composition is given by the following formulas:

$$m\left\{ Mg_3[Si_4O_{10}](OH)_2 \right\} \cdot p\left\{(Al, Fe)_2[Si_4O_{10}](OH)_2 \right\} \cdot nH_2O$$
$$\text{(montmorillonite)}$$

$$Al_2[Si_4O_{10}](OH)_2 \cdot nH_2O \quad \text{(beidellite)}$$

$$m\left\{ Mg_3[Si_4O_{10}](OH)_2 \right\} \cdot p\left\{(Fe, Al)_2[Si_4O_{10}](OH)_2 \right\} \cdot nH_2O$$
$$\text{(nontronite)}$$

$$Al_4[Si_4O_{10}](OH)_8 \quad \text{(kaolinite)}$$

$$Al_4[Si_4O_{10}] \cdot (OH)_2 \cdot 4H_2O \quad \text{(halloysite)}$$

$$MgAl_2[Si_4O_{10}] \cdot 4H_2O \cdot nH_2O \quad \text{(palygorskite)}$$

Easily recognizable by their brown and red colors, colloidal
and metacolloidal hydroxides of iron are typical products of acid
weathering. They occur in the form of nodules, films, or smudges.
Carrying positive charges, they sorb anions. Vanadium, phosphorus,
arsenic, uranium, molybdenum, indium, and boron have been

identified as impurities. The most important mineral of this group is limonite (hydrogoethite), $Fe_2O_3 \cdot H_2O \cdot nH_2O$.

Colloidal hydroxides of aluminum are considerably less common. They form under conditions of weathering and sedimentation in hot and humid climates and occasionally during the sulfatic weathering of rocks. Bauxites are mixtures of diaspore and boehmite, $Al_2O_3 \cdot H_2O$, and gibbsite, $Al_2O_3 \cdot 3H_2O$. Impurities encountered in bauxites include titanium, vanadium, gallium, strontium, columbium, tantalum, and, occasionally, boron and uranium.

The following metacolloids should be mentioned: delterite, $TiO_2 \cdot H_2O$; baddeleyite, ZrO_2; tungstite, $WO_3 \cdot H_2O$; "wood tin," SnO_2; navajoite, $V_2O_5 \cdot 3H_2O$; and nasturan, $kUO_2 \cdot hUO_3 \cdot mPbO$.

In addition, there are known colloidal and metacolloidal carbonates (e.g., calcite, smithsonite, and siderite), sulfates (barite), phosphates, and vanadates. Such minerals were discussed elsewhere (Chukhrov, 1955).

Chapter 6

Occurrence of Chemical Elements
in the Supergene Zone

The ability of a chemical element to migrate through the supergene zone depends on the manner in which it occurs, i. e., on whether it is in solution, incorporated in a crystal lattice, in a living matter, etc.

A knowledge of the modes of occurrence is essential to the understanding of such phenomena as the fertility of soils or the formation of dispersion haloes.

The principal modes recognized by Vernadskii were: (1) rocks and minerals, including water and gases; (2) magmas; (3) dispersions; and (4) living matter. In turn, the first of these groups may be subdivided into a number of smaller divisions. For example, iron may occur in the trivalent state when in feldspar lattices, in the sulfide form in pyrite, or as ferric hydroxide in limonite. Although the environment remains the same, each mode of iron will have a different mobility.

In oxidizing environments, trivalent iron migrates sluggishly out of limonite or feldspars. Pyritic iron migrates so energetically that sulfuric acid and ferrous sulfate form, the latter soluble in acidic water.

A chemical element occupying some definite place in a crystal lattice can go into solution only after that lattice is destroyed. On the other hand, an element situated in a "hole" in a crystal lattice can be removed easily. For example, uranium is easily leached out of igneous rocks by treatment with sodium carbonate solution.

Friedland (1955) classifies all components of soils as (1) active, (2) surface-active, (3) potential, or (4) inert. The first of these groups consists of water-soluble gases and exchange ions. The

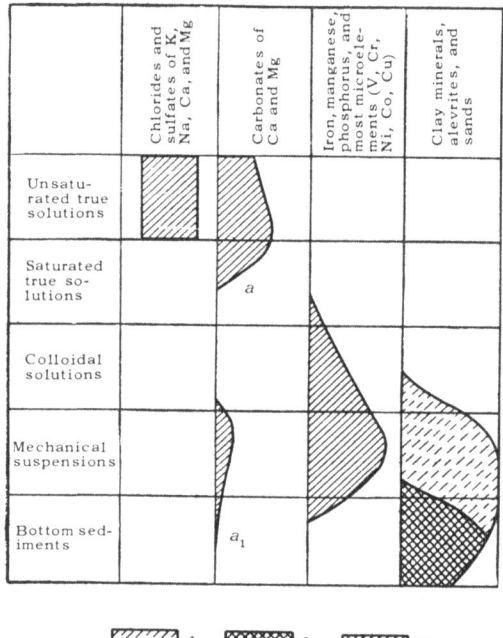

Fig. 18. Modes of stream transfer of principal components of sediments (after Strakhov, 1960). (1) Clay minerals; (2) minerals of sands and rock fragments; (3) other (a and a_1 are only for mountain creeks in semiarid climates.)

second group includes humus, clay minerals, and hydroxides. The third group consists of compounds which yield members of the other three groups when decomposed. The fourth group includes minerals which resist weathering, such as quartz and corundum.

There are two main modes of occurrence in water: (1) true (ionic) and (2) colloidal solution. For example, iron can occur in water as Fe^{2+}, Fe^{3+}, $Fe(OH)_2^+$, $Fe(OH)^{2+}$, and $Fe(OH)_3^\circ$.

Strakhov and his coworkers (1960) have shown that easily soluble salts are carried in streams in true solution, whereas iron, manganese, phosphorus, and rare elements are transported in suspension or as adsorbed on clay particles (Fig. 18). He gave the following generalization for the streams draining into the Black Sea:

1. Vanadium migrates only in suspension

2. Chromium, nickel, beryllium, gallium, and zinc migrate mostly in suspension

3. Iron, manganese, phosphorus, barium, copper, and strontium migrate both in suspension and in true solution, the tendency to migrate in solution increasing from iron to strontium

It is obvious that these generalizations may not hold for streams of another watershed.

The distinction between inert and mobile modes of occurrence has a definite practical significance.

The mobile mode of occurrence is that form of the element's occurrence in rocks, soils, or waters, which easily goes into solution and migrates.   Depending on conditions, the same form of an element may be inert in one environment and mobile in another. For example, ferric hydroxides are inert during weathering processes in deserts and semideserts, but are mobile in acid swamps of the subarctic.

An element is said to be inert if it occurs in conditions precluding its migration.

The mobility of chemical elements is evaluated in Table 19.

In geochemical exploration, the need often arises for a determination of not only the total content of a chemical element but also of the content of its mobile phase. To this end, its mobile phase is extracted from rocks by leaching with an aqueous, acid, neutral, alkaline, or some other appropriate solution. The method lends itself particularly well to the outlines of dispersion haloes. Dolukhanov's method of topological—hydrochemical surveying is one of several variants of this method.

Elements existing in soluble form, such as Na_2SO_4 and $NaCl$, are taken up in aqueous extracts. Gypsum can be only partly leached, while calcium carbonate is very poorly soluble in water. Generally speaking, aqueous extracts are analyzed to find the pH and determine the presence of chlorine, sulfate ion, carbonate ion, calcium, magnesium, sodium, and insoluble residue. The dried residue is further analyzed spectrographically. In the investigation of dispersion haloes, water-soluble compounds of ore elements are determined.

Insoluble carbonates of calcium and magnesium are among the less mobile modes of occurrence of those elements in rocks and soils. To determine their contents in a sample, the sample is treated with 2% HCl solution, which dissolves these carbonates without appreciably attacking the silicate phase of the sample.

On basis of this "carbonate" analysis, the contents of Ca, Mg,

Table 19. Principal Modes of Occurrence of Chemical Elements in the Epigenetic Zone

Mode of occurrence	Examples	
	Elements, compounds	Geological environment
Gaseous	O_2, N_2, CO_2, H_2S, CH_4, Rn, He, Ar	Surface and subsurface atmospheres, natural waters, living matter and some minerals (He in uraniferous, A in potash ones)
Soluble salts and their ions in solution	NaCl, Na_2SO_4, Na_2CO_3, $ZnSO_4$, $CuSO_4$, Na^+, Cu^{2+}	Soils, weathered crusts, terrigenous sediments, lakes of deserts and semideserts, oxidation zones of sulfide ore deposits. Ground waters of arid regions, deep-seated ground waters, brines, salt beds, hydrothermal solutions
Difficultly soluble	$CaCO_3$, $CaSO_4 \cdot 2H_2O$, $CuCO_3 \cdot Cu(OH)_2$, $PbSO_4$	Soils, weathered crusts and terrigenous sediments, deserts, and semideserts; partly in oxidation zones of sulfide ore deposits, ground waters, hydrothermal solutions
Elements in plants, animals, microorganisms	Albumens, fats, hydrocarbons, vitamins, and other organics consisting mainly of C, H, O, N, but also containing S, P, K, Ca, Mg, Cu, Zn, etc.	Landscapes of dry land, particularly in warm and humid climates (tropical forests) and to a lesser extent taiga (Siberian forest), tundra, steppes, deserts. Upper portions and shorelines of seas and oceans; also deep-seated ground waters (microorganisms)
Colloidal solutions and precipitates	Humates, colloidal precipitates of ferric hydroxide, silica, manganese, aluminum and, in part, clay minerals	Soil, weathered crusts and muds of water basins, hydrothermal solutions, and veins
Adsorbed ions — mostly cations	Ca^{2+}, Mg^{2+}, Na^+, Cu^{2+}, H, Ni^{2+}, Al^{3+}, etc., in colloidal minerals	Soils and muds of water basins, weathered crusts clayey sediments, iron and manganese

Table 19 (continued)

Mode of occurrence	Examples	
	Elements, compounds	Geological environment
		ores, carbonaceous — siliceous shales, coal, peats
Interphases in breaks in lattices	U	Rocks
Elements fixed in vertices and inter- sections of lattices of stable or fairly stable minerals	Si, Al, Zr, Hf, W, Sn, Ta, Nb, Th, rare earths	Zircon, sphene, quartz colum- bo — tantalates of soils weathered crusts, and terri- genous sediments

and CO_2 in the sample are calculated. This solution usually contains small amounts of the exchange calcium and magnesium of the noncarbonate form.

In carbonate-free rocks and soils, exchange cations are de- termined by a somewhat different technique involving a displace- ment of the given cation with the hydrogen ion of hydrochloric acid.

A soda carbonate leach is used to determine the elements which are soluble in alkaline media. Evseeva and Perel'man (1962) have extracted by this method some 90% of the total uranium content of a rock.

Concentrated acids, alone or in mixtures of two or more, are used to extract the less mobile forms of the chemical elements.

Natural Media of the Supergene Zone

The interaction of various geologic and climatic factors results in the formation of soils, crusts, weathered water-bearing strata, and other natural media.

SOIL

Soil, according to Vernadskii, is the region where living matter possesses the greatest amount of energy. It is the uppermost portion of the lithosphere, and participates in the biologic cycle by supporting vegetation. The process of photosynthesis ensures a continuous supply of organic matter to it. On decomposition, organic matter enriches the air and the water held in the soil in such energy carriers as CO_2, NH_3, etc.

Two types of soils are recognized by Polynov: eluvial and supraquatic.

Eluvial soils develop on water-divides and their slopes if the ground water table is very low. The migration of chemical elements in eluvial soils consists of two mutually opposed processes. Plants extract carbon dioxide from the air and their mineral salts extract it from the soil (Fig. 19).

Let us recall that plants possess the ability to sorb selectively. For example, the ashes of plants may contain up to several percent phosphorus, while the soils which feed those plants contain but a small fraction of one percent. Plants accumulate sulfur, calcium, potassium, phosphorus, copper, zinc, and many other elements.

Upon the death of plants and their subsequent decay, the upper layer of the soil becomes enriched in these elements, as was pointed out by the Norwegian geochemist Goldschmidt (1938).

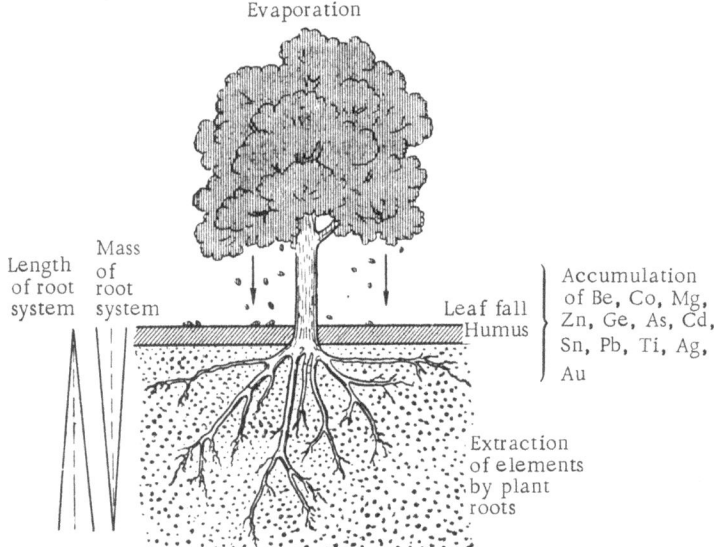

Fig. 19. Biogenic accumulation of elements in upper layers of soils (after Goldschmidt, 1938).

Figuratively speaking, plants "pump" certain chemical elements from the lower parts of the soil into the upper ones.

It was established by Dobritskaya et al. (1962) that the humic layers of podzol soils are enriched in Zn, Cu, Co, and Mo, while the same layers of chernozem or chestnut-brown soils are enriched only in Co and Cu. A similar accumulation of Zn and Co in gray forest and in alpine grassland-forest soils has been noted by Zhuravleva in the Chita state (1962).

Percolating downwards, atmospheric precipitation produces an opposite effect by transferring soluble compounds from the upper-most layers into the lower ones. Humic substances, including humates of heavy metals, are thus moved downwards, some pene-trating to the lower part of the soil and others moving still deeper.

Biogenic accumulation and leaching progress differently in different eluvial soils and either one may be predominant in the upper layers. As a result, different soil levels develop, varying widely in thickness. Each of these levels is characterized by its own peculiar chemical composition and its own physicochemical conditions. The upper levels may become acid while the lower become alkaline.

The total of various soil horizons present in a given soil is known as the soil profile. The structure of a soil and the distribution of metals in it may vary a great deal over its profile.

In fact, conditions may drastically change within a few centimeters.

The success or failure of a geochemical survey often depends on the correct recognition of this fact (Fig. 20).

The accumulation of humus and its mineral salts is characteristic of the uppermost level of most eluvian soils. In podzol soils, there is another level A_2 directly underneath A_1 in which the less soluble compounds accumulate while the more soluble ones are leached out.

Finally, most characteristic of eluvial soils is the level B where compounds which came from above accumulate. In chernozems, this B level is the place for the accumulation of calcium carbonate. In podzols and marsh soils, the organic and inorganic colloids may partially precipitate after having been leached out of the A levels. The illuvial B level gradually becomes clayey and may accumulate zinc, lead, and other metals which were sorbed elsewhere by the colloids (Fig. 21).

Supraquatic soils form wherever the ground water level is close to the surface — e.g., within a few meters. These soils are found in marshes, swamps, and grasslands. In addition to biogenic accumulation and leaching, chemical loads are also supplied by ground waters. Supraquatic soils can also be separated into layers. Swamp soils can be divided into layers of peat and inorganic colloids, and marshy soils into layers of gypsum and easily soluble salts.

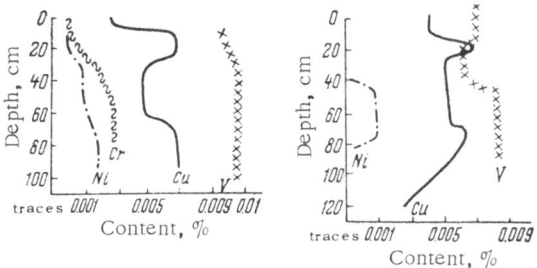

Fig. 20. Metal distribution curves along the vertical profile of dark gray forest soil of the southern Urals (after Bozhko, 1962).

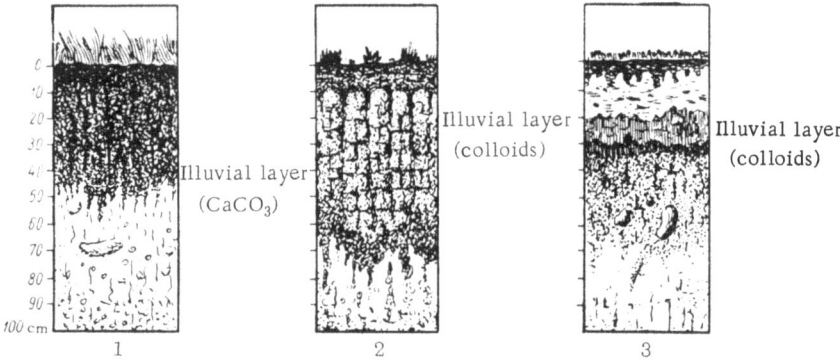

Fig. 21. Profiles of some soils: (1) chernozem, (2) caliche, (3) podzol.

To summarize, the most typical features of soil-forming processes are the biogenic accumulation of chemical elements in the uppermost layers and their differentiation in lower layers. Soils are rich in free energy, this accounting for the intense migration of chemical elements.

The distribution of metals in a soil profile should be correctly interpreted at the start of metallometric prospecting. Once found, the best layer alone should be sampled.

WEATHERED CRUST

The weathered crust (eluvium) is the name given to those products of rock alteration which accumulate residually. Although weathered rocks develop porous textures, they often retain some relict structures.* The contemporary weathered crust underlies eluvial soil and is formed by solutions descending from this soil. The immediate role of vegetation in its formation is not great, since vegetal roots penetrate only the upper layers.

A larger role in its formation is played by the products of the decomposition of organic matter — carbon dioxide and organic acids — which descend from above. Rocks are destroyed in the presence of such agents and are converted into porous, clayey

*Polynov (1934) distinguished between residual and the transported weathered crusts. We have just defined the first, while the second consists essentially of continental (terrigenous) sediments. As a rule, geologists accept the weathered crust to be synonymous with eluvium.

rubble. It is possible that certain microorganisms — so-called "silica bacteria" — aid in the formation of the weathered crust by destroying the silicate minerals.

The thickness of the crust varies from a fraction of one meter to several hundred meters, depending on such factors as climate, duration of the weathering, surface relief, tectonic regimen, and composition. The material composing the crust may be in any state: solid, liquid, or gas.

The optimum conditions for the formation of a thick crust are found in humid and hot climates combined with a gentle, rolling, nearly flat relief and a quiescent tectonic regimen (Fig. 22). In such cases, eluvium attains a thickness of several dozen meters, while alteration and fragmentation advance along fissures for hundreds of meters. In arid climates, weathering does not exceed more than a few meters. In mountainous regions, weathering becomes shallower as the hill slopes steepen.

The products of the weathering of ore outcrops differ radically from those of the enclosing rocks. The oxidation zone of sulfide ore is actually a special type of crust.

Like soils, weathered crusts display characteristic profiles which consist of a succession of layers of different mineral and chemical composition. The layers of crusts are generally thicker than those of soils.

Fig. 22. Kaolinitic, acid-weathered crust in humid tropics. The crust is thick and red. Colloids form in its lower part.

Fig. 23. Acid, hydromica crust in taiga (Siberian forest). The crust is brown and some-what less potent than that of the tropics. The bottom bleached zone is absent.

A fragmented horizon, representing a transition to fresh rock underneath, is common to all weathered crusts and develops at the foot of a cliff. Depending on the type of crust, this layer is covered with horizons of different composition: hydromicaceous, kaolinitic, ferruginous, aluminous, etc. In the arctic, fragmented horizons become thicker, especially at high altitudes (Figs. 23, 24, and 25).

Fig. 24. Carbonatic crust over basalts in semideserts. Because of $CaCO_3$, the crust is light-colored. It has a small thickness.

The behavior of chemical elements within the crust is conditioned by the following factors: (1) their chemical properties, (2) bioclimatic conditions, and (3) the mineral composition of the rocks. The ability to resist weathering varies with the species. For example, alkali amphiboles weather faster than albite. Therefore, it may be said that the sodium of alkali amphiboles possesses a greater migrational ability than the sodium of albite. The uranium contained in zircon has a lower migrational ability than the uranium of pitchblende.

The susceptibility of minerals toward chemical weathering has been appraised by various methods and is given in Table 20 in decreasing order of susceptibility.

Such data explain why quartz, rutile, tourmaline, or zircon are common, while olivine, pyrite, and feldspathoids are rare in weathering products. It should be kept in mind, however, that the data hold true only for humid climates. Calcite, dolomite, and gypsum are very stable in arid climates. Pyrite is very stable in subarctic environments.

The migration of chemical elements within the weathered crust involved both leaching and accumulation. The velocity of leaching varies over a very wide range. Chlorine and sulfur are removed several thousand times faster than aluminum, titanium, or iron. In general, it may be said that the formation of the weathered crust

Fig. 25. Clastic crust in high altitudes.

Table 20. Relative Stabilities of Minerals in Conditions of Chemical Weathering (after A. A. Koukharenko, G. Milner, etc.)

	Very stable	Stable	Rather stable	Unstable
Rock-Forming	Quartz	Muscovite Orthoclase Microcline Acid plagioclases	Amphiboles Pyroxenes Diopside-Heden- bergite	Basic plagioclases Feldspathoids Alkalic amphiboles Biotite Augite Olivine Glauconite Calcite Dolomite Gypsum
Accessory	Chromspinels Topaz Tourmaline Brookite Anatase Leucoxene Rutile Spinels Platinum Osmiridium Gold Zircon Iridium Corundum Diamond	Almandite Hematite Magnetite Titanomagnetite Columbite- Tantalite Sillimanite Kyanite Barite Thorianite Perofskite Ilmenite Xenotime Monazite Cassiterite Andalusite Garnets	Wolframite Scheelite Apatite Andradite Allanite Actinolite Zoisite Epidote Chloritoid Staurolite	Pyrrhotite Sphalerite Chalcopyrite Arsenopyrite Pyrite

consists of the removal of the more mobile components and the accumulation of the less mobile ones.

At the rate the crust loses its leachable components, it acquires other substances from the atmosphere — oxygen, water, and, occasionally, carbon dioxide, sulfur, and chlorine.

Occasionally, ore elements concentrate in commercial quantities in certain portions of the crust after having descended from above or having been brought in laterally by percolating waters.

A weathered crust is a region of intense oxidation. Having existed in igneous rocks in the bivalent state, iron, manganese, or sulfur oxidize to their highest valences during weathering.

Hydration is also typical in the crust. Most of the secondary minerals, including colloids, contain water of hydration, crystallization, etc., while most primary minerals are anhydrous.

Since secondary dispersion haloes, including buried ones, are often confined to the crust, a good knowledge of the geochemical peculiarities of the crust is necessary.

SURFACE AND GROUND WATERS

There are two processes controlling the chemical composition of waters. One of these develops in humid climates and inactive rocks, i.e., in rocks free from carbonates, gypsum, sulfides, or water-soluble salts. Much of the chemical load dissolved in water is contributed by organisms, whether these live in that water or are brought in by drainage. The decay of these organisms, i.e., their mineralization, imparts to water such loads as carbon dioxide, carbonate ion, calcium magnesium, phosphorus, sulfur, and humates. At the same time, the water is depleted of its free oxygen. The resulting chemical composition is fairly uniform and is not appreciably affected by regional structure.

In all arid climates, and also in some humid ones where rocks are chemically active, i. e., contain carbonates, sulfides, and solubles, the chemical composition of waters is mainly the result of the leaching of compounds contained in the surrounding rocks. Various metamorphic processes modify the composition of such waters, while evaporation is the principal means of raising concentrations. The composition is less likely to remain uniform or to be influenced by organic life.

GEOCHEMICAL LANDSCAPES

Geochemical landscapes are the result of the close interaction of many varied and complex factors involving soils, weathered crusts, continental sediments, surface and ground waters, and the adjacent atmospheric layers.

It is needless to say that macroclimatic and geologic conditions should remain uniform during the development of a given landscape.

The following examples illustrate the concept: (1) the granitoid hills of the Baikal region, (2) the forested chernozem steppes of the Ukraine, (3) the basalt plateaus of eastern Mongolia, (4) the desert sands of Karakum, (5) the permafrost regions of eastern Siberia, and (6) the karst topographies of the Urals.

The concept of geochemical landscapes was introduced by Polynov, who has drawn attention to the relationship between the conditions of the migration of chemical elements and configurations of the earth's surface. Landscapes are characterized by specific dispersion haloes.

WATER-BEARING STRATA OF THE CATAGENETIC SUBZONE

The alteration of rocks produced by ground waters is known as catagenesis. The region of catagenesis is the deepest portion of the supergene zone. In the catagenetic subzone, the epigenetic alteration of rocks generally takes place in two-phase systems (liquid + solid) and at higher temperatures and pressures. Water-bearing strata, and particularly their contacts with impermeable rocks, undergo the greatest alteration. Interiors of the water-impermeable strata often remain unaltered. Large blocks of unaltered rocks seem to have been isolated during the general alteration.*

Ground waters percolating through permeable rocks perform a large amount of chemical work — some minerals are precipitated, others are dissolved, still others are decrepitated. The composition of a water-bearing stratum changes substantially.

Like soils and weathered crusts, water-bearing strata are characterized by close relationships between their solid and liquid

---

*In 1922, Fersman wrote:
  "By catagenesis I mean that group of epigenetic processes which took place after the given stratum became covered with a new sediment and thus became separated from the water basin and prior to the uplift of the stratum and its exposure to the atmospheric air.... Catagenesis is due to the exchange of solutions between two strata which are different both chemically and petrographically. Its agents are oxygen, carbonic acid, water, and, to a lesser extent, silica, sulfatic solutions, etc."
Recently, the term catagenesis has been used in a broader sense to denote any alteration induced in rocks by ground waters. It has been so used by N. B. Vassoevich, G. I. Teodorovich, N. M. Strakhov, N. V. Logvinenko, A. I. Perel'man, etc.
  The formation of the soil and the formation of the weathered crust differ from catagenesis in that their changes are in three-phase systems (solid + liquid + gas).

phases. Therefore, their catagenesis should be investigated in all its various phases.*

Layers of epigenetically altered material develop as a result of catagenesis; we propose to call these "epigenetic horizons." Each such layer may encompass several strata or just a portion of one.

Until recently, the geochemical activity of ground waters was investigated by hydrologists who paid attention only to the waters themselves and not to the rocks altered by the waters. Still, many rocks now dry were water-bearing in the past. We have seen such manifestations — stringers of gypsum, silica, and other colloids — in the Urals, Kazakstan, Central Asia, etc.

Climate only slightly influences catagenetic processes. This influence is even smaller as depth increases. At greater depths, waters should be at a standstill in some aquifers and these should be free of any climatic influence. In a geologic cross-section, each type of water is represented by a specific coloration and a specific assemblage of secondary minerals.

Some catagenetic processes result in the formation of ore deposits and of the dispersion haloes around these ore deposits. The alternation of strata of different permeabilities is common. The ore deposits develop in the more permeable beds.

Depending on the movement of ground water, hydrogeologists subdivide the catagenetic subzone into three parts of the vertical hydrochemical zone (Table 21). Their order within the same water-bearing layer may be normal or inverted. Generally, this zone is well defined, although in places it might have been destroyed. It varies from one region to another. For example, carbon dioxide and hydrogen sulfide form during the oxidation of the organic matter contained in bituminous rocks, and, dissolving in water, impart to it a high chemical activity.† The upper, oxygenated subzone overlies the intermediate one, which contains H_2S and is characterized by negative Eh values. The lowest subzone contains hydrocarbons and possesses strongly negative Eh values.

---

*Some catagenetic alteration may be the result of the action of ooze water on the underlying rocks. Such cases arise during marine transgressions when new marine sediments begin to form over the already consolidated strata. Waters percolating from the ooze down into the underlying strata start various reactions of a reducing nature. Undoubtedly, such waters should be considered ground waters although they cannot be identified with any specific water-bearing stratum. An example will be discussed in Chapter 12.

†More about this in Chapter 12.

Table 21. A Scheme of Environmental Conditions of Formation of Ground Waters and Their Zonation (after N. K. Ignatovich)

Hydrodynamic zones		Genetic types of	Hydrogeochemical zones
Zone of active water regimen and underground drainage	Geologic and hypsometric allocation of zones	waters and types of ground water resources	Types of processes
Zone of active water exchange – surface drainages	Actively bathed parts of structures; zone of influence of stream network often extends to depths of 300 m or more	Modern meteoric waters are in motion. Their dynamic resources predominate over static	Active leaching of chlorides and sulfates decreasing from elevations to depressions and also toward arid climates
Zone of obstructed circulation of ground waters and slowed exchange	Deeper portions of flowing artesian basins, reaching to 500 m on platforms and to 2000 m in regions of folding and tectonic movements; thermal waters	Slowly renovating waters which replace more ancient ones. Static resources predominate over dynamic ones	Slow flushing from rocks of their salt complexes (marine, lagoon, and other faces). In course of long periods of time these waters change their profiles
Zone of stagnant water regimen. Underground drainage becomes noticeable only in geologically long periods of time	Usually in zone of deep-seated sedimentary complex but may lie closer to surface in closed structures	Ancient connate waters often in huge quantities	Accumulation of salts and metamorphism of waters; geochemical processes (diffusion, osmosis, etc.)

| Hydrogeochemical zones | | Examples | |
Development of chemical types of waters	Economic significance	Artesian basin of Moscow	Northern Caucasus
In depressions in arid regions, mostly fresh and bicarbonatic and also sulfatic or chlorsulfatic waters	Predominantly fresh waters for potable, livestock, and industrial uses	Ground waters of Carboniferous, Mesozoic, and Quaternary strata — spheres of active exchange — drainages of upper Volga, Oka, Moscow, or Klyaz'ma Rivers. Fresh, bicarbonatic waters	Waters of Mesozoic strata in highlands: predominantly fresh or highly mineralized waters
In regions of young folding: bicarbonatic; also thermal or chlorsulfatic, alkaline, or H_2S waters	Mainly chlorsulfatic or alkaline; thermal or with indications of oil deposits being destroyed: medicinal or industrial uses	U. Devonian waters in feeble circulation. Waters intermediate between bicarbonatic and sulfatic or chlorsulfatic waters	Waters in Mesozoic and L. Tertiary strata of the foothills. Obstructed circulation in deep tectonic breaks. Composition: between alkalisulfatic and chlorsulfatic
Highly mineralized waters and brines of the Cl—Na—Ca type	Industrial brines with B, I, Ra, and other rare elements. Oil deposits	M. Devonian brines, highly mineralized and metamorphized to a large degree. The Cl—Na—Ca type	Waters of Cretaceous and Tertiary strata in places of transition from folded regions to platforms. The Cl—Na—Ca type of waters

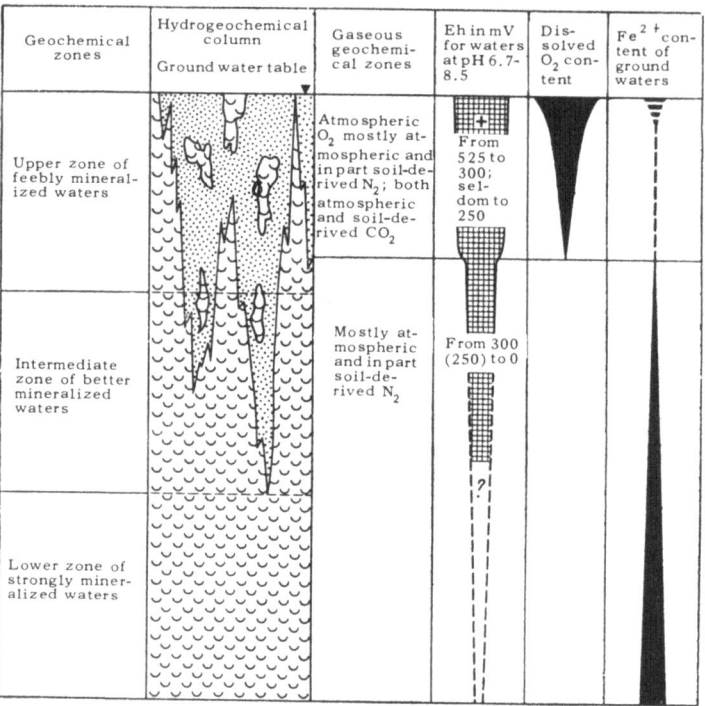

Geochemical zones	Hydrogeochemical column — Ground water table	Gaseous geochemical zones	Eh in mV for waters at pH 6.7-8.5	Dissolved O_2 content	Fe^{2+} content of ground waters
Upper zone of feebly mineralized waters		Atmospheric O_2 mostly atmospheric and in part soil-derived N_2; both atmospheric and soil-derived CO_2	From 525 to 300; seldom to 250		
Intermediate zone of better mineralized waters		Mostly atmospheric and in part soil-derived N_2	From 300 (250) to 0		
Lower zone of strongly mineralized waters					

Fig. 26. Hydrochemical zoning in a layer lacking reactive organic matter (after Germanov et al., 1962).

The nature of the water zone is entirely different in areas where rocks are free from organic matter — e.g., volcanics. Hydrogen sulfide is absent. The Eh values are positive even at considerable depths where free oxygen is also lacking.

Rukhin (1961) has detected a very definite zoning of epigenetic processes which corresponds to the above hydrochemical zoning. His stage of the removal of uncombined water from the rocks corresponds to the zone of active water circulation; his zone of the removal of weakly bonded water corresponds to the zone of obstructed circulation; and his zone of the removal of firmly bonded water corresponds to the zone of stagnation. Densities and other properties of rocks change accordingly. According to Babinets (1961), the composition of ground waters from deeper portions of the Ukrainian artesian basin is similar to that of pore water contained in shales. In fact, he believes that most of the water in deep artesian basins has been derived by the filter-pressing of water from consolidating clay sediments.

See Figs. 26 and 27 for the processes taking place in zones of active and partly obstructed circulation. Epigenetic alteration in deeper regions and at higher temperatures and pressures are omitted, since it borders on metamorphism.

WEATHERING AND CEMENTATION

It has been shown that rocks and waters interact at approximately the same thermodynamic conditions in all natural media of the supergene zone. This accounts for certain peculiarities of physicochemical migration, which consists of diametrically opposed processes. First of all, a comminution of rocks and minerals takes place and leads to a decrease of density, an increase of porosity, and an increased dispersion of the chemical elements. Aquatic migrants are removed. Taken together, these processes constitute weathering. To be sure, some matter is also accumulated. The accumulating matter consists mainly of aerial migrants: carbon dioxide, oxygen, water, etc. The corresponding stages of weathering are carbonatization, oxidation, hydration, etc.

The opposing group of epigenetic processes is instrumental in the accumulation of aquatic migrants. The precipitation of metals out of solution, the cementation of rocks, a decrease in porosity, and an increase in density are the results. The entire strata may become cemented, or disconnected concretions may form, the cementing materials being ferric hydroxides, calcium carbonate, gypsum, etc. No specific name was given to this group of processes. Rather inadequately they are referred to as leaching and cementation.

Actually, weathering and cementation are two different phases of the same process. They develop side by side in any natural medium of the supergene zone, although not necessarily equally. In all eluvial soils, leaching and cementation result in the differentiation of calcareous, gypsiferous, ferruginous, and other strata. Cementation is very common in marshes.

Although weathering predominates within the crust, some cementation goes on in its lower portions, where silica concentrates and new clay minerals are formed.

Weathering is relatively slight in water-bearing strata and disappears entirely with depth. Nevertheless, some minerals become altered and even destroyed by ground water. Such phenom-

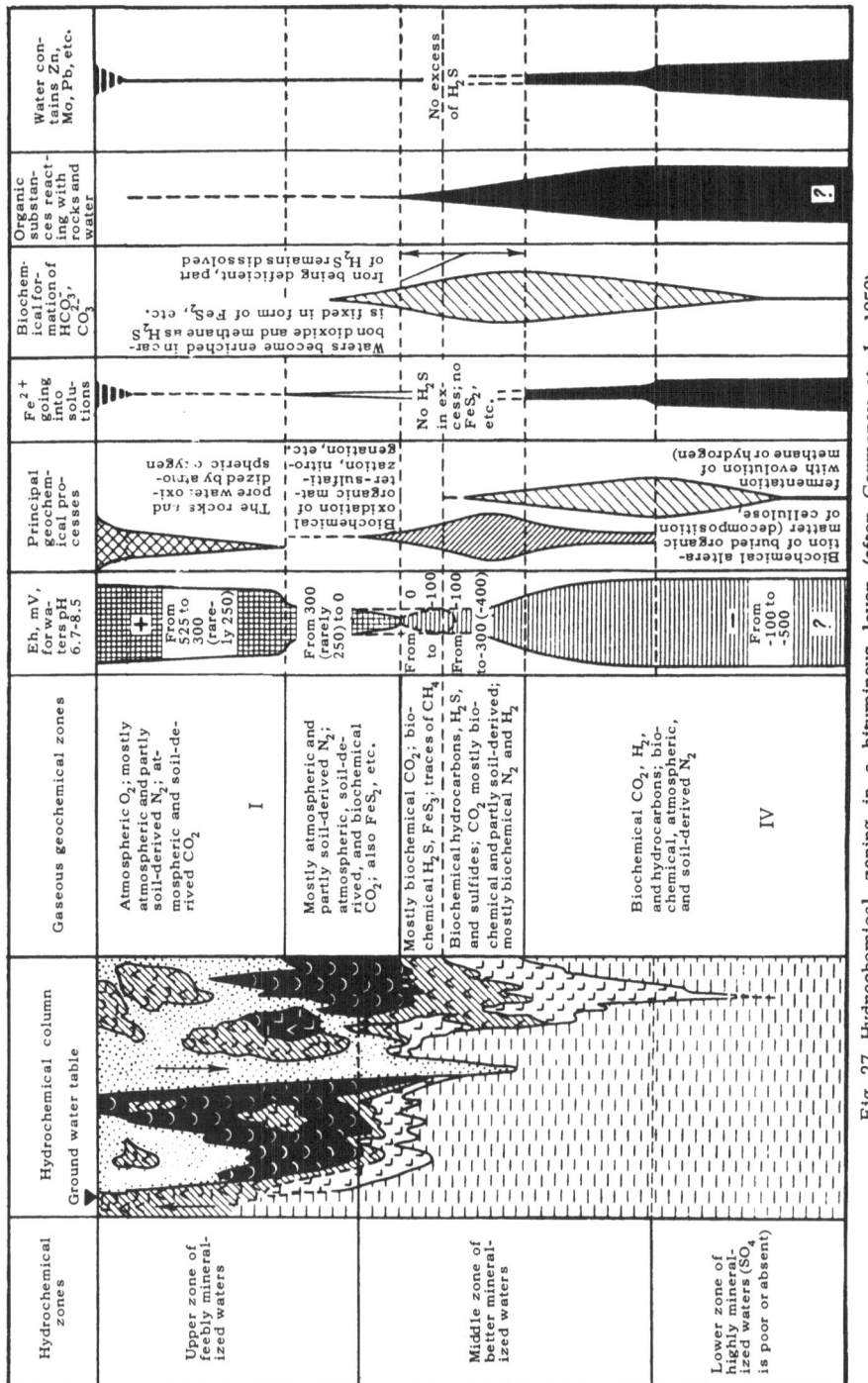

Fig. 27. Hydrochemical zoning in a bituminous layer (after Germanov et al., 1959).

ena are sometimes referred to as the "weathering in depth." However, the processes of leaching, cementation, and recrystallization are common enough in all water-bearing strata and intensify with depth. Shales may become cupriferous and uranium ores may form in such a manner.

EPIGENETIC ZONING

By epigenetic zoning we mean the paragenetic sequence of zones or layers which are formed in natural media by means of some epigenetic processes. Atoms migrate freely across such zones. Substances leached out of one zone may be redeposited in another. In most cases, their continuity is due to the differentiation of a single flow of water.

Epigenetic zoning in soils and weathered crusts was recognized long ago as the so-called "soil layers." Recently the investigation has been extended into the catagenetic subzone. Best studied are the interactions between oxygenated.waters and rocks containing some reducing agents such as pyritic shales, coal seams, and bituminous limestones. Zoning due to such oxidation will be discussed in Chapter 13.

In soils, the thickness of individual layers does not exceed 1 m, the total thickness of all layers being less than 2 m. In weathered crusts, the horizons may be several meters thick, with the total extending for hundreds of meters. In the catagenetic subzone, epigenetic zones may swell locally to hundreds of meters and even kilometers.

Chapter 8

Factors in the Development of Media
in the Supergene Zone

Within the supergene zone, the properties of natural media and the course of epigenetic processes are affected by two factors: the climate and the geologic structure.

CLIMATE

Climate affects many geochemical features of epigenetic processes, particularly in the uppermost portion of the lithosphere (soils and weathered crusts). Temperature and atmospheric precipitation are the most important climate elements.

In regions with humid climates, natural waters are actually unsaturated solutions of such elements as copper, selenium, etc. No equilibria are ever attained between such solutions and the rocks. Therefore, waters are aggressive and their chemical composition depends on the velocities of weathering and leaching, i.e., it depends on the duration of the interaction between water and rocks.

In arid climates, concentrations are raised by evaporation (Chapter 10) and waters may become saturated with respect to a number of components, such as Ca^{2+}, Mg^{2+}, HCO_3^-, and SiO_2. The composition of water depends upon the composition of rocks. The same waters remain undersaturated with respect to rare elements.

All this indicates the role climate plays in epigenetic processes. Climate also conditions the development of living matter, and this makes its influence on epigenetic processes so much greater. Let us recall that solar energy is the energy source for all

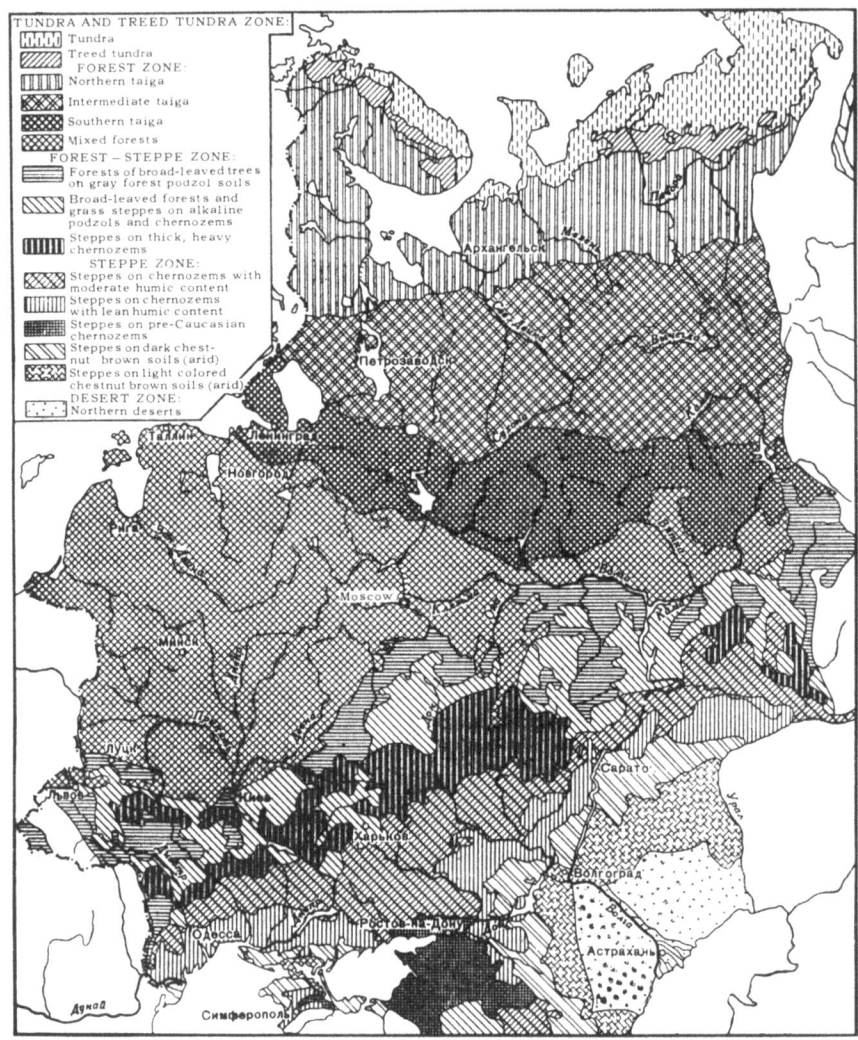

Fig. 28. Landscape zones and subzones of plains of the European USSR.

epigenetic processes, to which it is supplied indirectly in the course
of the geologic cycling of atoms. When released from a cycle into
an epigenetic process, this energy is usually in a form capable of
doing some chemical work.

In warmer and more humid climates more organic matter is
created. The more this organic matter is decomposed, the more
carbon dioxide, organic acids, and other compounds will be supplied

to waters.   As a result, the epigenetic processes will develop more vigorously.

In regions with a warm, humid climate and a flat topography, the migration of elements is both biogenic and physicochemical in nature.   In the arctic, in deserts, or on high plateaus, the mechanical mode of migration is predominant.

Although the effects of climate are best noted at the earth's surface, even artesian waters show some such influence. According to Frolov (1963), the thermal regimen of rocks shows a climatic zoning down to a depth of 3 km.

ZONING

Climatic zones are generally followed by zones of soils, crusts, terrigenous sediments, ground water, etc.

Alexander Humboldt's concept that vegetation follows climatic zones has been extended by Dokuchaev to cover minerals. Detailed studies show that vegetation, animal life, soils, weathered crusts, terrigenous sediments, and even concretions are distributed zonally. In turn, all these affect the landscape zoning.

On the plains of the western USSR, various landscape zones and subzones succeed each other northwest to southeast in the following climatic order (Figs. 28 and 29):

1. Glaciated zone
2. Zone of arctic deserts
3. Tundra zone
   a. Northern tundra (moss and lichens)
   b. Middle tundra
   c. Southern tundra
   d. Forested tundra
4. Taiga (forest) zone
   a. Northern taiga
   b. Middle taiga
   c. Southern taiga
   d. Subzone of mixed forests
5. Zone of forested steppes
6. Zone of steppes
   a. Subzone of northern chernozem steppes
   b. Subzone of southern chernozem steppes
   c. Subzone of northern steppes with dark chestnut-brown
      soils

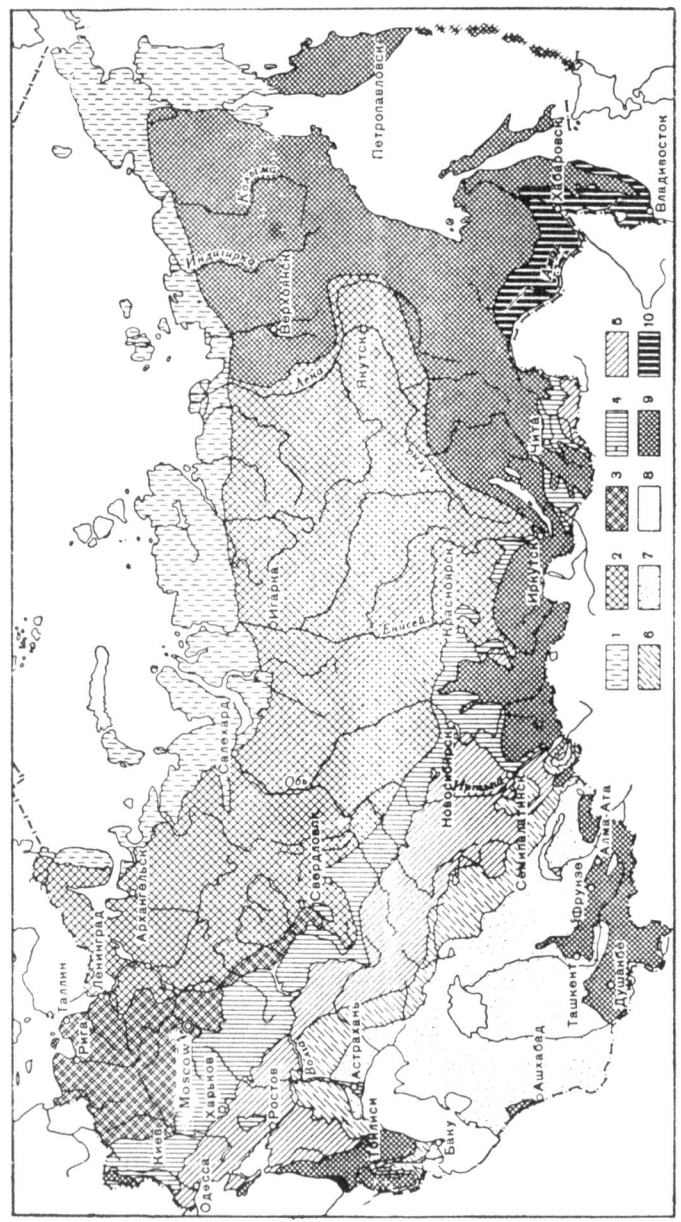

Fig. 29. Landscape zones of the USSR (after Berg). (1) Tundra; (2) taiga; (3) mixed forest; (4) transition from forest to steppe; (5) steppe; (6) semidesert; (7) desert; (8) subtropics; (9) highland landscapes; (10) broad-leaved and mixed forests of the Far East.

      d. Subzone of southern steppes with light chestnut-brown
         steppes
7. Zone of deserts
      a. Subzone of northern deserts (Kazakhstan type)
      b. Subzone of southern deserts (Central Asiatic type)

Each of these landscape zones is a definite geochemical zone characterized by a specific biologic cycling of atoms and a specific mode of the aqueous transfer of atoms. As with climates, these zones may be further subdivided into subzones which differ mainly in the intensities of the epigenetic processes. For example, weathering is slower in the northern taiga than it is in the southern.

The vertical succession of the zones is observable in mountainous regions. Thus, steppes developed at the foot of the northern Caucasus change upwards into forests (foliated at first, then evergreen), then into alpine meadows, and finally into landscapes with naked cliffs and talus. The highest peaks are covered with snow and eternal ice (Fig. 30).

The vertical zoning of landscapes changes in accordance with the geographical position of the mountain range and its height. For example, the only zones present in the northern Urals or in some mountain regions of Siberia are taiga (the lower zone) and alpine meadows (the upper zone). In the mountains of Central Asia, the forest zone may be absent and the steppes may grade directly into the alpine meadows.

The vertical and the horizontal zoning are not analogous. Some zones found in the mountains have no counterparts on the plains. The subzone of alpine meadows is an example.

While many zones originated in climatic zoning, the same zones in different media have different dimensions. This is due to the second independent variable—the geologic factor. It is independent of modern climates.

The closer a given medium lies to the surface of the earth the more it is influenced by the climate. The influence of the geologic structure increases with depth from the surface. Vegetation, which is mostly dependent on the climate and not so much on the structure, displays the most marked zoning. The geographic distribution of soils, weathered crusts, and the terrigenous sediments follow climatic boundaries. Individual zones in the crusts are generally wider that the corresponding zones in soils or vegetation. Several soil or vegetation zones may fit within the limits

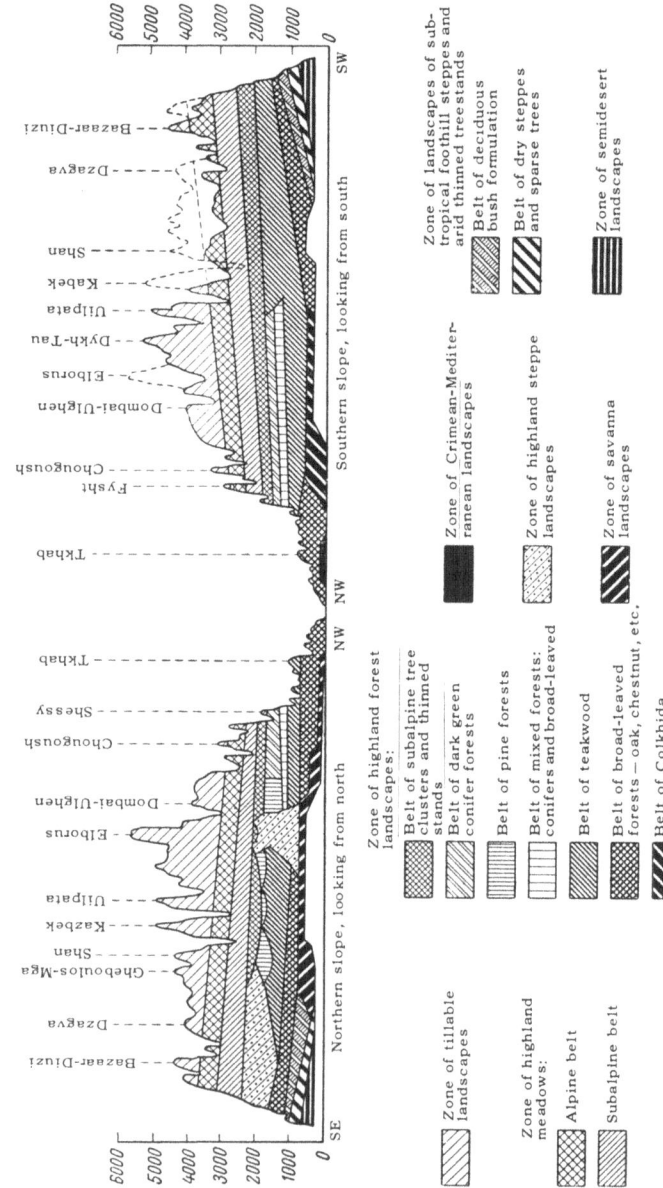

Fig. 30. A scheme of landscape zones and belts of the Greater Caucasus (after Kachashkina, Geographical Museum, University of Moscow).

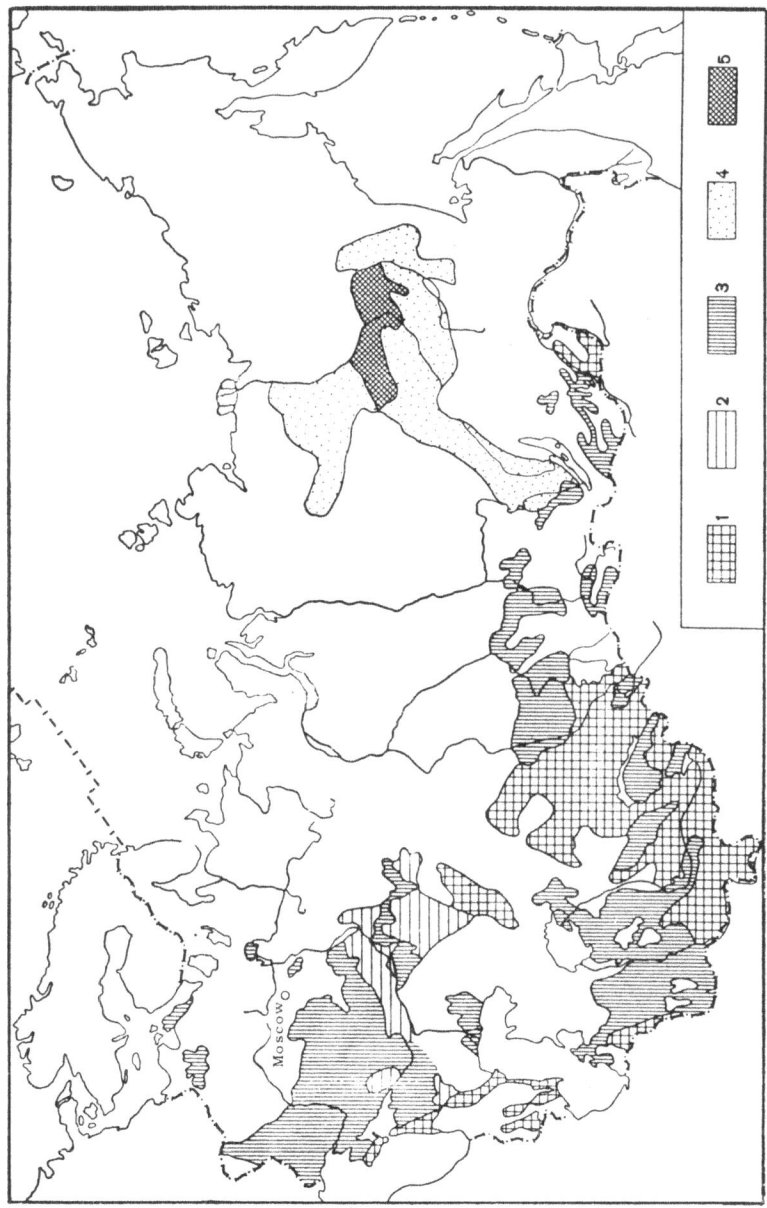

Fig. 31. Occurrences of the carbonatic weathered crust. (1) Carbonatic ortho- and para-eluvium, locally in combination with clastic crust; (2) carbonatic para-eluvium; (3) carbonatic neo-eluvium; (4) carbonatic para-eluvium in combination with carbonatic clastic and locally also with acid colloform or acid crusts; (5) carbonatic neo-eluvium.

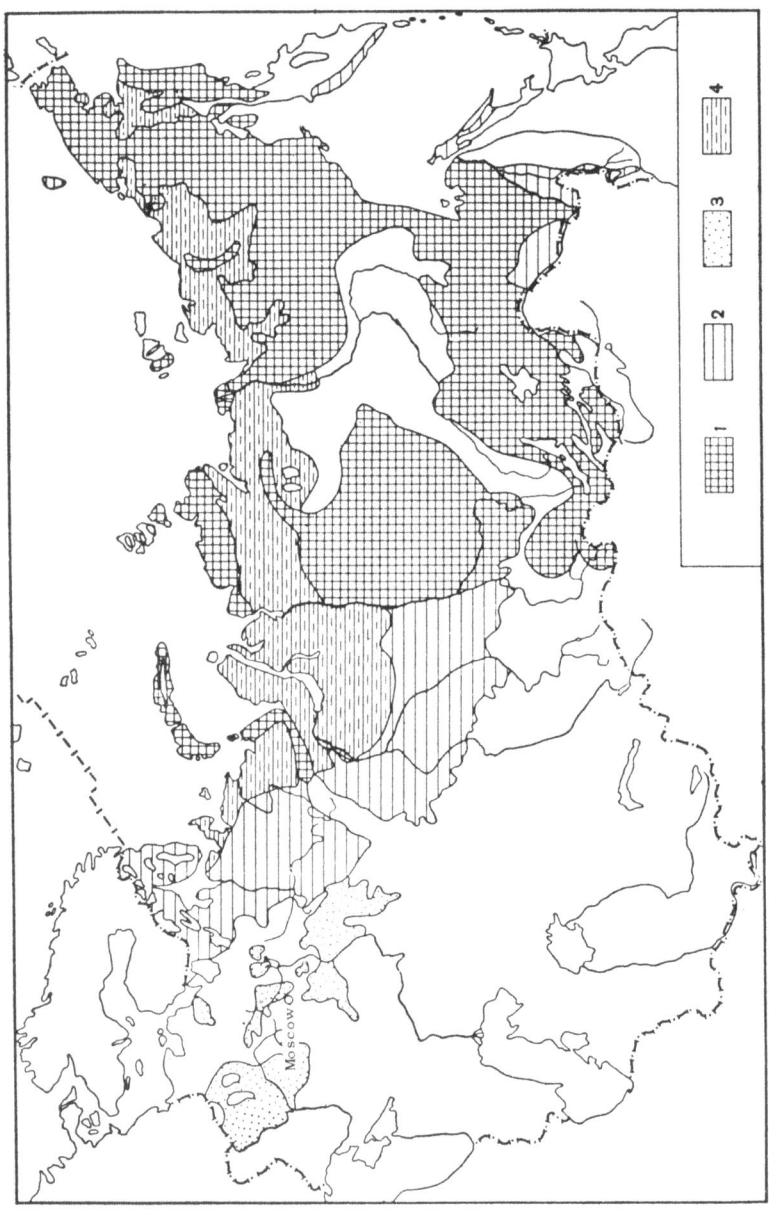

Fig. 32. Occurrences of acid colloform crust. (1) Acid colloform ortho- and para-eluvium in combination with acid clastic crust; (2) acid colloform neo-eluvium; (3) acid colloform neo-eluvium in combination with acid crust; (4) acid colloform neo-eluvium in regions of permafrost.

of a single zone in the crust. For example, the single, carbonatic zone of the crust corresponds to the combined zones of chernozem, chestnut-brown, brown, and gray soils. These, in turn, correspond to many vegetation zones of northern steppes, typical steppes, dry steppes, and deserts. Within the limits of the single acid colloidal crust there can develop three soils—the tundra soil, the podzol, and the grassy-podzol (Figs. 31 and 32).

The zoning of ground waters may become locally enhanced by various geologic conditions (Fig. 33). Various manifestations of zoning can also be distinguished: stream waters, lacustrine sediments, etc.

GEOLOGIC STRUCTURE

The role of geologic structure in epigenetic processes increases with depth from the surface. The role is slight in soils, but increases in the weathered crust or in ground waters and becomes very important in deeper parts of the lithosphere. However, the geologic factor may overshadow the importance of climate. For example, the oxidation of sulfidic ore bodies is confined to places where such bodies occur and is modified by the relationship existing between the ore and the enclosing host rocks. The effect of climate is secondary in importance since sulfuric acid is generated in dry and humid climates alike. In humid climates, this sulfatic type is very distinct from the usual lateritization. In arid climates, it also differs from the usual carbonatization.

Still, the climate does exert a definite influence on the geographic distribution of the weathered crusts. It imposes a pattern of zoning against the background of such geologic conditions as regional folding, or sulfidic mineralization.

The composition of rocks is reflected in the processes taking place in soils or other media. Climatic conditions remaining the same, the same epigenetic process will develop differently in limestones, pyritic shale, or salt-impregnated shales, etc.

S. S. Smirnov (1955) divided the minerals occurring in the oxidizing sulfide ores into active, semiactive, and inactive classes. Obviously, the rocks containing active minerals would be oxidizable to the greatest extent. Rocks containing active minerals react vigorously with the percolating waters.

The concept of mineral activity can be extended to other environ-

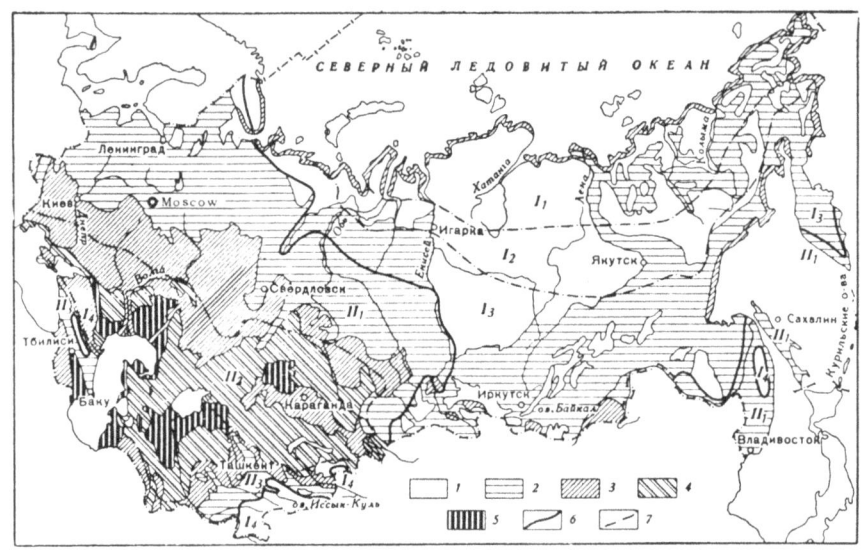

Fig. 33. Ground water zones in the area of free water circulation of the USSR (after Zaitsev and Rasponov). Mineralization and chemical composition of waters in soil and subsoil: (1) Very sweet; with mineralization up to 0.2 g/liter; predominantly bicarbonatic but often carrying large amounts of silica or organic matter; locally in maritime regions with sulfatic or chloridic contents. (2) Sweet, bicarbonatic waters; with mineralization up to 0.5 g/liter. (3) Sweet to feebly saline with mineralization generally up to 1 g/liter and locally up to 3 g/liter; mostly bicarbonatic but in southern regions with sulfatic or chloridic contents. (4) Sweet to saline with mineralization up to 10 g/liter or even more; usually sulfatic and/or chloridic but occasionally bicarbonatic. (5) From sweet to brine with mineralization up to 200 g/liter and even more; predominantly chloridic, less often sulfatic and sodium-bicarbonatic; in well-flushed regions sweet to slightly saline and then mostly bicarbonatic and/or sulfatic. (6) Boundary between permafrost and climatic provinces. (7) Boundary between belts (zones).

Ground waters in stable perennial permafrost: (I_1) Belt of perennial permafrost with a few thaws and widely developed fossil ice, with a total thickness of frozen rock of 200 to 500 m and more. (I_2) Belt of perrenial permafrost with widely developed thaws; total thickness of permafrosted rocks 100 to 200 m (in Viliuisk syneclyse, 400 to 600 m). (I_3) Belt of "islands" of permafrosted areas with thicknesses of permafrosted rocks of 25 to 100 m or even less. (I_4) Belt of alpine glaciers and snows.

Ground waters in provinces free from permafrost: (II_1) Belt of predominant leaching of salts; humates-rich. (II_2) Arid belt of continental accumulation in rocks and waters. (II_3) Belt of vertically zoned processes of leaching and accumulation in mountainous regions.

Table 22. Chemical Activity of Minerals Compared to That
of Hydrogen Ion (after Udodov et al., 1959)

Mineral	pH values		Activity class
	Acid medium, $AK = 3.0$	Water, $AV = 6.2$	
Marcasite	2.85	2.75	I
Pyrrhotite	3.0	5.75	
Pyrite	3.0	5.8	
Diaspore	3.0	6.2	
Rhodochrosite	3.0	6.2	
Stibnite	3.0	6.2	
Boulangerite	3.0	6.2	II
Galena	3.0	6.3	
Orpiment	3.0	6.3	
Magnetite	3.0	6.4	
Albite	3.0	6.5	
Jarosite	3.0	6.7	
Siderite	3.0	6.8	
Fluorite	3.08	6.2	
Diopside	3.1	6.2	
Serpentine	3.1	6.2	
Sphalerite	3.1	6.4	
Hematite	3.1	6.5	
Olivine	3.1	6.7	
Orthoclase	3.1	7.25	
Hornblende	3.1	7.4	III
Labradorite	3.1	7.8	
Aegirine-Augite	3.1	7.9	
Actinolite	3.1	8.15	
Nepheline	3.2	6.3	
Cerussite	3.2	6.4	
Barite	3.2	7.2	
Chromite	3.25	7.4	
Smithsonite	3.3	6.4	
Wollastonite	3.4	6.6	
Magnesite	3.5	8.6	
Anglesite	4.2	6.2	
Rhodonite	3.65	7.2	IV
Gypsum	4.4	6.8	
Apatite	4.8	8.7	
Muscovite	5.35	8.9	
Dolomite	5.5	8.9	
Iceland Spar	8.1	9.5	V
Calcite	8.2	9.2	

Table 23. Rock Classification Chart Based on Comparison of Their
Chemical Activities (after Udodov, 1959)

Activity class	Rock type	Rocks	Number of samples	AK = 30	AV = 6.2
I	Active	Eruptive, metamorphic and sedimentary rocks with iron sulfides	15	Under 3	Under 6.2
II	Neutral	Eruptive, metamorphic and sedimentary rocks; unconsolidated nonloess sediments	50	3.0	6.2—6.4
III	Very feebly active	Nonstratified eruptive metamorphic and sedimentary rocks — mainly with alkaline earths	130	3.1—3.4	6.5—7.0
IV	Intensified activity due to secondary activity	Fissured extrusive, metamorphic and sedimentary rocks with fissures filled with calcite and calcareous cement. Unconsolidated sediments approaching the loess types	90	3.5—6.0	7.1—8.0
V	Strongly active precipitants of heavy metals	Massive sedimentary carbonatites with Ca and alkalies (marble, limestone, dolomite, etc). Unconsolidated, loess-like sandy clays	75	Over 5.0	8.1—9.5

ments of the supergene zone.   A mineral may be active in one medium and inactive in another.   For example, sulfides of heavy metals, pyrite included, are active in the weathered crust, where waters are oxygenated and oxidizing toward those sulfides. The same minerals would be inactive at depths where waters lack oxygen.   Calcite is inactive in humid climates where ground water is weakly mineralized and has an acid reaction favorable to the leaching of limestones.   In arid climates, it is usually inactive, since the strongly mineralized chlor-sulfatic waters are already saturated with calcium carbonate.

Udodov, Onufrienok, and Kristalev (1959) have introduced the concept of the chemical activity of rocks.   This activity is measured by the change in pH of a solution as it interacts with rocks and minerals.   These authors suggested that the chemical activity of rocks be measured at pH 3 for acid solution (AK) and at pH 6.2 for distilled water (AV). The data measured under such conditions are given in Tables 22 and 23.*

Yurkevich (1958) introduced the concept of reducing capacity, RC, which is determined by the amount of oxygen needed to oxidize reducing minerals such as FeS_2 and $FeCO_3$. His data indicate that 1 to 13 mg O_2 are needed for every 100 g of sulfide-bearing Mesozoic clays of western Siberia. The RC factor is less than unity for igneous rocks and in some cases it equals zero. Rocks with a high RC factor affect the percolating ground waters and thus affect various supergene processes. For example, some epigenetic deposits of uranium are formed in such a manner (Chapter 13).

The technique of RC determination has been worked out by Bardoshi and Bod (1960) and by Evseeva (Evseeva and Perel'man, 1962). In general, a sample is leached in a chromic mixture and the pH of the liquid is determined. All such methods have only a relative merit, since the natural processes are much more complex and are affected by biochemical factors.

It is needless to say that the chemical composition of rocks is not the only influencing factor. Tectonics, stratigraphy, incumbent volcanism, and many other phenomena should be taken into account.

---

*To determine AK, a sample is crushed to 0.1–0.25 mm and then treated with a standard solution having a pH of 3 (0.001 N H_2SO_4 in 2% Na_2SO_4). The mixture is centrifuged and the pH of the liquid phase is read. To determine AV, the sample is treated with distilled water at a pH of 6.2.

Chapter 9

Classification of the Epigenetic Processes
Operating in the Supergene Zone

Although epigenetic processes are numerous, they can be grouped into a few types on the basis of their typomorphic constituents.

TYPOMORPHIC ELEMENTS

Chemical elements, ions, and compounds are said to be typomorphic if their migration characterizes a given epigenetic process. In turn, typomorphism depends on the abundance and the migrational ability of an element.

With regard to their abundance, chemical elements are either principal or accessory. To be principal, the elements should be abundant enough to constitute the bulk of rocks: Si, Al, Fe, Ca, Mg, Na, K, P, Cl, Ti, etc. Being present in significant amounts, such elements control the migration of other elements. Some of the principal elements might be typomorphic.

Elements which normally occur only in minor amounts are said to be accessory (Table 24). Such elements do not exert any influence on the environment. Consequently, they cannot be typomorphic. For example, sodium, constituting 2.5% of the lithosphere, is a typomorphic element in a number of supergene processes — such as the formation of salt beds and the migration of silica. On the other hand, cesium, rubidium, and lithium, although analogous to sodium, cannot be typomorphic because of their extremely small abundances. If their abundances had been as great as that of sodium, their roles in supergene processes would have been equally

Table 24. The Average Abundance of Chemical Elements in the Lithosphere and in Principal Rock Types (after Vinogradov, 1962), %

Ele-ment	Ultra-basic	Basic	Inter-mediate	Acidic	Clays, shales, slates, schists	Average litho-sphere
H	—	—	—	—	—	—
Li	$5 \cdot 10^{-5}$	$1.5 \cdot 10^{-3}$	$2 \cdot 10^{-3}$	$4 \cdot 10^{-3}$	$6 \cdot 10^{-3}$	$3.2 \cdot 10^{-3}$
Be	$2 \cdot 10^{-5}$	$4 \cdot 10^{-5}$	$1.8 \cdot 10^{-4}$	$5.5 \cdot 10^{-4}$	$3 \cdot 10^{-4}$	$3.8 \cdot 10^{-4}$
B	$1 \cdot 10^{-4}$	$5 \cdot 10^{-4}$	$1.5 \cdot 10^{-3}$	$1.5 \cdot 10^{-3}$	$1 \cdot 10^{-2}$	$1.2 \cdot 10^{-3}$
C	$1 \cdot 10^{-2}$	$1 \cdot 10^{-2}$	$2 \cdot 10^{-2}$	$3 \cdot 10^{-2}$	1.0	$2.3 \cdot 10^{-3}$
N	$6 \cdot 10^{-4}$	$1.8 \cdot 10^{-3}$	$2.2 \cdot 10^{-3}$	$2 \cdot 10^{-3}$	$6 \cdot 10^{-2}$	$1.9 \cdot 10^{-3}$
O	42.5	43.5	46.0	48.7	52.8	47.0
F	$1 \cdot 10^{-2}$	$3.7 \cdot 10^{-2}$	$5 \cdot 10^{-2}$	$8 \cdot 10^{-2}$	$5 \cdot 10^{-2}$	$6.6 \cdot 10^{-2}$
Na	$5.7 \cdot 10^{-1}$	1.94	3.0	2.77	0.66	2.50
Mg	25.9	4.50	2.18	0.56	1.34	1.87
Al	0.45	8.76	8.85	7.70	10.45	8.05
Si	19.0	24.0	26.0	32.3	23.8	29.5
P	$1.7 \cdot 10^{-2}$	$1.4 \cdot 10^{-1}$	$1.6 \cdot 10^{-1}$	$7 \cdot 10^{-2}$	$7.7 \cdot 10^{-2}$	$9.3 \cdot 10^{-2}$
S	$1 \cdot 10^{-2}$	$3 \cdot 10^{-2}$	$2 \cdot 10^{-2}$	$4 \cdot 10^{-2}$	$3 \cdot 10^{-1}$	$4.7 \cdot 10^{-2}$
Cl	$5 \cdot 10^{-3}$	$5 \cdot 10^{-3}$	$1 \cdot 10^{-2}$	$2.4 \cdot 10^{-2}$	$1.6 \cdot 10^{-2}$	$1.7 \cdot 10^{-2}$
K	$3 \cdot 10^{-2}$	$8.3 \cdot 10^{-1}$	2.3	3.34	2.28	2.50
Ca	0.70	6.72	4.65	1.58	2.53	2.96
Sc	$5 \cdot 10^{-4}$	$2.4 \cdot 10^{-3}$	$2.5 \cdot 10^{-4}$	$3 \cdot 10^{-4}$	$1 \cdot 10^{-3}$	$1 \cdot 10^{-3}$
Ti	$3 \cdot 10^{-2}$	$9 \cdot 10^{-1}$	$8 \cdot 10^{-1}$	$2.3 \cdot 10^{-1}$	$4.5 \cdot 10^{-1}$	$4.5 \cdot 10^{-1}$
V	$4 \cdot 10^{-3}$	$2 \cdot 10^{-2}$	$1 \cdot 10^{-2}$	$4 \cdot 10^{-3}$	$1.3 \cdot 10^{-2}$	$9 \cdot 10^{-3}$
Cr	$2 \cdot 10^{-1}$	$2 \cdot 10^{-2}$	$5 \cdot 10^{-3}$	$2.5 \cdot 10^{-3}$	$1 \cdot 10^{-2}$	$8.3 \cdot 10^{-3}$
Mn	$1.5 \cdot 10^{-1}$	$2 \cdot 10^{-1}$	$1.2 \cdot 10^{-1}$	$6 \cdot 10^{-2}$	$6.7 \cdot 10^{-2}$	$1 \cdot 10^{-1}$
Fe	9.85	8.56	5.85	2.70	3.33	4.65
Co	$2 \cdot 10^{-2}$	$4.5 \cdot 10^{-3}$	$1 \cdot 10^{-3}$	$5 \cdot 10^{-4}$	$2 \cdot 10^{-3}$	$1.8 \cdot 10^{-3}$
Ni	$2 \cdot 10^{-1}$	$1.6 \cdot 10^{-2}$	$5.5 \cdot 10^{-3}$	$8 \cdot 10^{-3}$	$9.5 \cdot 10^{-3}$	$5.8 \cdot 10^{-3}$
Cu	$2 \cdot 10^{-3}$	$1 \cdot 10^{-2}$	$3.5 \cdot 10^{-3}$	$2 \cdot 10^{-3}$	$5.7 \cdot 10^{-3}$	$4.7 \cdot 10^{-3}$
Zn	$3 \cdot 10^{-3}$	$1.3 \cdot 10^{-2}$	$7.2 \cdot 10^{-3}$	$6 \cdot 10^{-3}$	$8 \cdot 10^{-3}$	$8.3 \cdot 10^{-3}$
Ga	$2 \cdot 10^{-4}$	$1.8 \cdot 10^{-3}$	$2 \cdot 10^{-3}$	$2 \cdot 10^{-3}$	$3 \cdot 10^{-3}$	$1.9 \cdot 10^{-3}$
Ge	$1 \cdot 10^{-4}$	$1.5 \cdot 10^{-4}$	$1.5 \cdot 10^{-4}$	$1.4 \cdot 10^{-4}$	$2 \cdot 10^{-4}$	$1.4 \cdot 10^{-4}$
As	$5 \cdot 10^{-5}$	$2 \cdot 10^{-4}$	$2.4 \cdot 10^{-4}$	$1.5 \cdot 10^{-4}$	$6.6 \cdot 10^{-4}$	$1.7 \cdot 10^{-4}$
Se	$5 \cdot 10^{-6}$	$5 \cdot 10^{-6}$	$5 \cdot 10^{-6}$	$5 \cdot 10^{-6}$	$6 \cdot 10^{-5}$	$5 \cdot 10^{-6}$
Br	$5 \cdot 10^{-5}$	$3 \cdot 10^{-4}$	$4.5 \cdot 10^{-4}$	$1.7 \cdot 10^{-4}$	$6 \cdot 10^{-4}$	$2.1 \cdot 10^{-4}$
Rb	$2 \cdot 10^{-4}$	$4.5 \cdot 10^{-3}$	$1 \cdot 10^{-2}$	$2 \cdot 10^{-2}$	$2 \cdot 10^{-2}$	$1.5 \cdot 10^{-2}$
Sr	$1 \cdot 10^{-3}$	$4.4 \cdot 10^{-2}$	$8 \cdot 10^{-2}$	$3 \cdot 10^{-2}$	$4.5 \cdot 10^{-2}$	$3.4 \cdot 10^{-2}$
Y	—	$2 \cdot 10^{-3}$	—	$3.4 \cdot 10^{-3}$	$3 \cdot 10^{-3}$	$2.9 \cdot 10^{-3}$
Zr	$3 \cdot 10^{-3}$	$1 \cdot 10^{-2}$	$2.6 \cdot 10^{-2}$	$2 \cdot 10^{-2}$	$2 \cdot 10^{-2}$	$1.7 \cdot 10^{-2}$
Nb	$1 \cdot 10^{-4}$	$2 \cdot 10^{-3}$	$2 \cdot 10^{-3}$	$2 \cdot 10^{-3}$	$2 \cdot 10^{-3}$	$2 \cdot 10^{-3}$
Mo	$2 \cdot 10^{-5}$	$1.4 \cdot 10^{-4}$	$9 \cdot 10^{-5}$	$1 \cdot 10^{-4}$	$2 \cdot 10^{-4}$	$1.1 \cdot 10^{-4}$
Ru	—	—	—	—	—	—
Rh	—	—	—	—	—	—
Pd	$1.2 \cdot 10^{-5}$	$1.9 \cdot 10^{-6}$	—	$1 \cdot 10^{-6}$	—	$1.3 \cdot 10^{-6}$
Ag	$5 \cdot 10^{-6}$	$1 \cdot 10^{-5}$	$7 \cdot 10^{-6}$	$5 \cdot 10^{-6}$	$1 \cdot 10^{-5}$	$7 \cdot 10^{-6}$
Cd	$5 \cdot 10^{-6}$	$1.9 \cdot 10^{-5}$	—	$1 \cdot 10^{-5}$	$3 \cdot 10^{-5}$	$1.3 \cdot 10^{-5}$
In	$1.3 \cdot 10^{-6}$	$2.2 \cdot 10^{-5}$	—	$2.6 \cdot 10^{-5}$	$5 \cdot 10^{-6}$	$2.5 \cdot 10^{-5}$

Table 24 (continued)

Element	Ultra-basic	Basic	Inter-mediate	Acidic	Clays, shales, slates, schists	Average litho-sphere
Sn	$5 \cdot 10^{-5}$	$1.5 \cdot 10^{-4}$	—	$3 \cdot 10^{-4}$	$1 \cdot 10^{-3}$	$2.5 \cdot 10^{-4}$
Sb	$1 \cdot 10^{-5}$	$1 \cdot 10^{-4}$	$2 \cdot 10^{-5}$	$2.6 \cdot 10^{-5}$	$2 \cdot 10^{-4}$	$5 \cdot 10^{-5}$
Te	$1 \cdot 10^{-7}$	$1 \cdot 10^{-7}$	$1 \cdot 10^{-7}$	$1 \cdot 10^{-7}$	$1 \cdot 10^{-6}$	$1 \cdot 10^{-7}$
I	$1 \cdot 10^{-6}$	$5 \cdot 10^{-5}$	$3 \cdot 10^{-5}$	$4 \cdot 10^{-5}$	$1 \cdot 10^{-4}$	$4 \cdot 10^{-5}$
Cs	$1 \cdot 10^{-5}$	$1 \cdot 10^{-4}$	—	$5 \cdot 10^{-4}$	$1.2 \cdot 10^{-3}$	$3.7 \cdot 10^{-4}$
Ba	$1 \cdot 10^{-4}$	$3 \cdot 10^{-2}$	$6.5 \cdot 10^{-2}$	$8.3 \cdot 10^{-2}$	$8 \cdot 10^{-2}$	$6.5 \cdot 10^{-2}$
La	—	$2.7 \cdot 10^{-3}$	—	$6 \cdot 10^{-3}$	$4 \cdot 10^{-3}$	$2.9 \cdot 10^{-3}$
Ce	—	$4.5 \cdot 10^{-4}$	—	$1 \cdot 10^{-2}$	$5 \cdot 10^{-3}$	$7 \cdot 10^{-3}$
Pr	—	$4 \cdot 10^{-4}$	—	$1.2 \cdot 10^{-3}$	$5 \cdot 10^{-4}$	$9 \cdot 10^{-4}$
Nd	—	$2 \cdot 10^{-3}$	—	$4.6 \cdot 10^{-3}$	$2.3 \cdot 10^{-3}$	$3.7 \cdot 10^{-3}$
Pm	—	—	—	—	—	—
Sm	—	$5 \cdot 10^{-4}$	—	$9 \cdot 10^{-4}$	$6.5 \cdot 10^{-4}$	$8 \cdot 10^{-4}$
Eu	$1 \cdot 10^{-6}$	$1 \cdot 10^{-4}$	—	$1.5 \cdot 10^{-4}$	$1 \cdot 10^{-4}$	$1.3 \cdot 10^{-4}$
Gd	—	$5 \cdot 10^{-4}$	—	$9 \cdot 10^{-4}$	$6.5 \cdot 10^{-4}$	$8 \cdot 10^{-4}$
Tb	—	$8 \cdot 10^{-5}$	—	$2.5 \cdot 10^{-4}$	$9 \cdot 10^{-5}$	$4.3 \cdot 10^{-4}$
Dy	$5 \cdot 10^{-6}$	$2 \cdot 10^{-4}$	—	$6.7 \cdot 10^{-4}$	$4.5 \cdot 10^{-4}$	$5 \cdot 10^{-4}$
Ho	—	$1 \cdot 10^{-4}$	—	$2 \cdot 10^{-4}$	$1 \cdot 10^{-4}$	$1.7 \cdot 10^{-4}$
Er	—	$2 \cdot 10^{-4}$	—	$4 \cdot 10^{-4}$	$2.5 \cdot 10^{-4}$	$3.3 \cdot 10^{-4}$
Tu	—	$2 \cdot 10^{-5}$	—	$3 \cdot 10^{-5}$	$2.5 \cdot 10^{-5}$	$2.7 \cdot 10^{-5}$
Yb	—	$2 \cdot 10^{-4}$	—	$4 \cdot 10^{-4}$	$3 \cdot 10^{-4}$	$3.3 \cdot 10^{-5}$
Lu	—	$6 \cdot 10^{-5}$	—	$1 \cdot 10^{-4}$	$7 \cdot 10^{-5}$	$8 \cdot 10^{-5}$
Hf	$1 \cdot 10^{-5}$	$1 \cdot 10^{-4}$	$1 \cdot 10^{-4}$	$1 \cdot 10^{-4}$	$6 \cdot 10^{-4}$	$1 \cdot 10^{-4}$
Ta	$1.8 \cdot 10^{-6}$	$4.8 \cdot 10^{-5}$	$7 \cdot 10^{-5}$	$3.5 \cdot 10^{-4}$	$3.5 \cdot 10^{-4}$	$2.5 \cdot 10^{-4}$
W	$1 \cdot 10^{-5}$	$1 \cdot 10^{-4}$	$1 \cdot 10^{-4}$	$1.5 \cdot 10^{-4}$	$2 \cdot 10^{-4}$	$1.3 \cdot 10^{-4}$
Re	—	$7.1 \cdot 10^{-8}$	—	$6.7 \cdot 10^{-8}$	—	$7 \cdot 10^{-8}$
Os	—	—	—	—	—	—
Ir	—	—	—	$6.3 \cdot 10^{-7}$	—	—
Pt	$2 \cdot 10^{-5}$	$1 \cdot 10^{-5}$	—	—	—	—
Au	$5 \cdot 10^{-7}$	$4 \cdot 10^{-7}$	—	$4.5 \cdot 10^{-7}$	$1 \cdot 10^{-7}$	$4.3 \cdot 10^{-7}$
Hg	$1 \cdot 10^{-6}$	$9 \cdot 10^{-6}$	—	$8 \cdot 10^{-6}$	$4 \cdot 10^{-5}$	$8.3 \cdot 10^{-6}$
Tl	$1 \cdot 10^{-6}$	$2 \cdot 10^{-5}$	$5 \cdot 10^{-5}$	$1.5 \cdot 10^{-4}$	$1 \cdot 10^{-4}$	$1 \cdot 10^{-4}$
Pb	$1 \cdot 10^{-5}$	$8 \cdot 10^{-4}$	$1.5 \cdot 10^{-3}$	$2 \cdot 10^{-3}$	$2 \cdot 10^{-3}$	$1.6 \cdot 10^{-3}$
Bi	$1 \cdot 10^{-7}$	$7 \cdot 10^{-7}$	$1 \cdot 10^{-6}$	$1 \cdot 10^{-6}$	$1 \cdot 10^{-6}$	$9 \cdot 10^{-7}$
Po	—	—	—	—	—	—
Rn	—	—	—	—	—	—
Ra	—	—	—	—	—	—
Ac	—	—	—	—	—	—
Th	$5 \cdot 10^{-7}$	$3 \cdot 10^{-4}$	$7 \cdot 10^{-4}$	$1.8 \cdot 10^{-3}$	$1.1 \cdot 10^{-3}$	$1.3 \cdot 10^{-3}$
Pa	—	—	—	—	—	—
U	$3 \cdot 10^{-7}$	$5 \cdot 10^{-5}$	$1.8 \cdot 10^{-4}$	$3.5 \cdot 10^{-4}$	$3.2 \cdot 10^{-4}$	$2.5 \cdot 10^{-4}$

Note: Lithosphere = 1 part basic + 2 parts acidic.

great.   Calcium, sulfur, and iron are typomorphic in proper environments while radium, tellurium, or cobalt never are.

The dividing line between principal and accessory elements is
very arbitrary. Generally speaking, the element's influence on the
geochemistry of supergenesis is negligible if its content in rocks
is less than 0.01%. Should the content of the same rare element rise
locally, as in sulfide ore deposits, its role increases and may even
become typomorphic.

As already stated, not all principal elements can be typomorphic. The greatest role is played by those elements which are
capable of both active migration and residual concentration. Thus,
sluggish titanium exerts very little influence on epigenetic processes in spite of its relatively high abundance. This leads us to
formulate the following principle of elemental mobility: Typomorphic elements are those which migrate the most actively
during a given process, or accumulate residually in the greatest
quantities (Perel'man, 1955).

AEOLEAN AND AQUEOUS MIGRANTS

Two main types of typomorphic elements can be distinguished.
Aeolean typomorphic elements and compounds
migrate in the gas state. Examples include carbon dioxide, hydrogen sulfide, and methane. Their influence is great, as was first
recognized by Vernadskii (1933) and later expounded by Ovchinnikov
(1955) in his geochemical classification of ground waters.

Gases mainly influence oxidation–reduction reactions. Three
principal environments can be distinguished in this connection: (1)
oxidizing, (2) reducing in the absence of H_2S, and (3) reducing in the
presence of H_2S.

Oxidizing environments are characterized by the presence of
free oxygen or some other strong oxidizers dissolved in water.
Iron, manganese, copper, vanadium, sulfur, and many other elements are encountered here in their highest valences. Rocks are
colored red, brown, or yellow. In alkaline media, the Eh is only
slightly above zero, usually above 0.15 V and sometimes up to
0.7 V. In acidic media, the lower limit of the oxidizing environment displays an Eh above 0.4 V, while the Eh interval of 0.15 to
0.4 V corresponds to the reducing conditions. Oxygen is the typomorphic element here.

Reducing environments, lacking hydrogen sulfide, are characterized by waters which contain little or no free oxygen or other strong oxidizers. Locally they contain much carbon dioxide or methane. Iron and manganese migrate easily in the bivalent form. The typomorphic gases here are carbon dioxide and (locally) methane. The Eh values are below 0.4 V in acidic media and below 0.15 V in alkaline. Rocks are colored greenish, gray, or bluish gray. At considerable potential drops, vanadium and copper are reduced: V^{5+} to V^{3+}, and Cu^{2+} to Cu^+ and further to $Cu°$. Soluble compounds of these metals are formed during this reduction.

Waters of the remaining category of reducing environments, those with abundant hydrogen sulfide, contain large quantities of this gas and sometimes methane and other hydrocarbons but no oxygen or other strong oxidizers. Forming insoluble sulfides, iron and a number of other heavy metals do not migrate. The typomorphic compounds here are hydrogen sulfide and occasionally hydrocarbons. This environment is noted for its predominantly alkaline conditions, with pH values over 7. The Eh frequently falls below zero and as low as −0.6 V. The difference between this and the preceding environments lies in the H_2S content and not in the redox potential. The Eh remaining the same and the H_2S content varying, the environments change from the second to the third type, e.g., from taiga swamps to desert marshes.

Water-transported typomorphic elements and compounds are those which characteristically migrate in true and colloidal solutions. Examples are chloride ion, sulfate ion, bicarbonate ion, calcium, magnesium, and sodium. These migrants exert a considerable influence on supergene processes and often impose certain peculiar aspects on these.

In the environmental series of aqueous migration, each environment is characterized by some typomorphic ions and compounds. These control the alkalinity–acidity of natural waters and their chemical loads. The most important of these environments are:

1. Strongly acidic. The pH is less than 4. The commonest typomorphic ion is H^+; occasionally typomorphic are Fe^{3+}, Al^{3+}, Zn^{2+}, Cu^{2+}, etc.
2. Acidic. The pH varies between 4 and 6.5. Typomorphic ions are H^+ and anions of organic acids.
3. Neutral to weakly alkaline, enriched in calcium bicarbonate. The pH varies between 6.5 and 8.5. Typomorphic ions are Ca^{2+}, HCO_3^-, etc.

4. Neutral and weakly alkaline, chlor-sulfatic. The pH varies between 7 and 8. The typomorphic ions are Na^+, Cl^-, SO_4^{2-}.
5. Neutral and weakly alkaline, gypsum-depositing. The typomorphic ions are SO_4^{2-} and Ca^{2+}.
6. Alkaline, sodic. The pH is above 8.5. Typomorphic ions and compounds are HCO_3^-, OH^-, Na^+, SiO_2, etc.

INTENSITY OF AQUEOUS MIGRATION
OF CHEMICAL ELEMENTS

A complete description of the behavior of the chemical elements includes an appraisal of the intensity of their aqueous migration. The data of ordinary chemical analyses do not adequately portray the idea because of differences in the abundances of those elements. For example, it cannot be predicted from such data (Table 25) whether silicon or zinc migrates most energetically. While silicon appears to predominate over zinc in water analyses (10 mg/liter Si and 0.05 mg/liter Zn), the average content of silicon in rocks is also larger than that of zinc (29.5% Si and 0.008% Zn).

Quantitatively, the intensity of aqueous migration of an element is expressed by the equation of migrational ability (Perel'man, 1956, 1961). Assuming the quantity of the element X in a given natural system to be b_x, its amount in the mobile state at any given moment of time is:

$$\frac{\Delta b_x}{b_x} \cdot \frac{1}{\Delta t}.$$

Table 25. Calculation of the Coefficient of Aqueous Migration K_x from the Abundance of an Element in Stream Water m_x, and in the Lithosphere n_x

Abundance	Units	Si	Ca	Zn	Cu	Fe
Stream waters, m_X	mg/liter	10	50	$5 \cdot 10^{-2}$	$3 \cdot 10^{-3}$	1
Lithosphere, n_X	%	29.5	2.96	$8.3 \cdot 10^{-3}$	$4.7 \cdot 10^{-3}$	4.65
Coefficient, K_X	—	0.07	3.3	1.2	0.12	0.04

Note: The mineral residue of water is assumed to weigh 500 mg/liter; the content of the elements in rocks is taken at their abundance values.

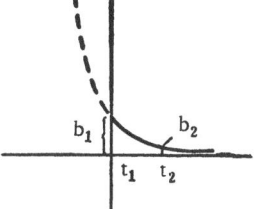

Fig. 34. A decrease in the content of a chemical element with time in the weathered crust, the rate of change assumed to be constant.

In its application to the weathered crust, this quantity, P_x, expresses the velocity of leaching of 1 g of a substance containing a given element. Rewriting it in the differential form and noting that $db/b = d \ln b$, we get

$$P_x = \frac{d \ln b_x}{dt}.$$

Unfortunately, the equation has no numerical solution, since the rate of change of the quantity of the given element in rocks with time is unknown (i.e., $b = f(t)$ is unknown).

If the intensity P_x remains constant during a given process, the solution of the equation is simple:

$$P_x \cdot dt = d \ln b_x,$$

$$\int_{t_1}^{t_2} P_x dt = \int_{b_1}^{b_2} d \ln b_x,$$

$$P_x (t_2 - t_1) = \ln \frac{b_2}{b_1},$$

$$P_x = \frac{\ln \frac{b_2}{b_1}}{t_2 - t_1},$$

and

$$b_2 = b_1 \cdot e^{-P_x (t_2 - t_1)}.$$

In other words, the decrease of the amount of a substance of a given element during leaching is an exponential function (Fig. 34).

We have calculated the values of ($\ln b_2 - \ln b_1$) for various elements and various conditions of weathering and found it to vary within the limits of n and 0.0 (Perel'man, 1941, 1956, etc.). For example, the following figures were obtained for the lateritic weathered crust in the Batum region (Perel'man and Batulin, 1962):

$$SiO_2 = -1.58 \qquad CaO = -3.8$$

$$TiO_2 = -2.73 \qquad MgO = -0.75$$
$$Al_2O_3 = -1.09 \qquad K_2O = -4.37$$
$$Na_2O = -3.08$$

Assuming that weathering continued in that region during the entire Quaternary Era (one million years) and that the intensity of leaching remained the same, the fractions leached yearly per 1 g of the substance containing the given element are:

$$SiO_2 = -1.58 \cdot 10^{-6} \qquad CaO = -3.08 \cdot 10^{-6}$$
$$TiO_2 = -2.73 \cdot 10^{-6} \qquad MgO = -7.5 \cdot 10^{-7} \text{ etc.}$$

Approximate as these figures are, they give an idea of the intensities of aqueous migration.

Since the determination of P_x is quite difficult in most cases, the author has proposed the use of the coefficient of aqueous migration, K_x, which is obtained by dividing the element's content in the rock into its content in waters draining from that rock.* The greater the value of K_x, the greater will be the migrational ability of a given element:

$$K_x = \frac{m_x 100}{an_x},$$

where m_x is the X element's content in water, mg/liter, n_x is its content in the rock, %, and a is the mineral residue contained in the water, %.

The K_x factors afford a convenient means of comparing the intensities of migration of the principal and accessory elements. Thus, returning to the example of silicon and zinc, we note that zinc is 17 times more mobile than silicon (Table 25). These factors can be used for other purposes, such as the speed of leaching in the weathered crust or in the catagenetic zone. In the last case it is advisable to use the lithospheric abundance of an element for its n_x because an unknown number of rocks could have contributed to the chemical composition of water in deep-seated strata.

Elements may be arranged in series in accordance with their K_x factors (Table 26).

*The following relationship exists between the values of P_x and K_x:

$$\frac{P_x}{P_y} = \frac{K_x}{K_y}$$

according to Perel'man (1956).

Table 26. Migrational Series during the Weathering of Siliceous Rocks in Temperate Climates (following Polynov)

Migrational intensity	Oxidizing environment K_x 1000 100 10 1 0.1 0.01 0.001	Contrast of migration ← Weak ⌐ Strong	Strongly reducing environments K_x 1000 100 10 1 0.1 0.01 0.001
Very strong	Cl, I Br, S	Cl, Br, I	Cl, I Br
Strong	Ca, Mg Na, F Sr, Zn U	Ca, Mg, Na, F, Sr / Zn, U	Ca, Mg Na, F, Sr
Medium	Ca, Si, P, Cu Ni, Mn, K	Si, P, K / Cu, Ni, Co	Si, P, K
Weak and very weak	Fe, Al, Ti, Y, Th, Zr, Hf, Nb, Ta, Ru, Rh, Pd, Os, Pt, Sn	Al, Ti, Zr, Hf, Nb, Ta, Pt, TR, Sn	Al, Ti, Sc, V, Cu, Ni, Co, Mo, Th, Zr, Hf, Nb, Ta, Ru, Rh, Pd, Os, Zn, U, Pt

During weathering, chlorine and sulfate sulfur are very easily leached out. The so-called "easily leachable elements," Ca, Mg, Na, F, and Sr stand next in the series, their K_x values varying between 20 and 1. Then follow the "mobile elements," Cu, Ni, Co, Mn, P, and silicate SiO_2, with K_x values fluctuating between 2 and 0.1. Finally, Fe, Al, Ti, rare earths, Zr, Hf, Pt, and other "inert and least mobile" elements end the series, with K_x values lower than 0.1.

A detailed study discloses that the elements move into water in groups characterized by certain ranges of velocity: Co, Ni, Cu, is one group, for example. The least mobile obviously accumulate residually.

It will be shown in the chapters which follow that the formation of the secondary dispersion haloes or even of some supergene ore bodies depends directly on the mode of migration. It should be said in this connection that the series of Table 26 hold true only in temperate climates. Even then, the K_x factors will be different for the weathering of limestones or rocks mineralized with sulfides or impregnated with salt. Thus, iron and aluminum acquire high mobilities in zones of oxidation of sulfide ore deposits. Weathering in tropical and polar regions again changes the K_x factors — iron, for example, becomes highly mobile in tundra.

A serious error incurred in the determination of K_x arises from the contamination by chemical elements of the atmosphere in seaside regions; Cl, Na, SO_4^{2-}, and even iodine can be contaminants. Therefore, the value of K_x is largely illustrative.

Every environment is characterized by its own migrational series. A comparison of the two environments in Table 26 — the oxidizing and the strongly reducing (rich in H_2S) — indicates the differences existing in the migrational abilities of the same element. This leads to the concept of "migrational contrasts." This contrast is measured by the ratio of the element's migrational intensities during the two different processes. For example, the contrast coefficient of zinc is nearly 100, which is very high. It is obtained by dividing the K_{zn} for the sulfide medium $(0.0n)$ into the K_{zn} for the oxide medium (n).

For elements like Zr, Hf, T, and Pt, which do not readily form soluble compounds in the supergene zone, the K_x does not exceed 0.01. It follows that all inert metals have low contrast coefficients.

In a way, the contrast coefficients depend also on the media which are being compared. In general, an element with a higher contrast is more likely to concentrate in the form of ores.

The classification scheme for chemical elements which follows is based on (1) the mode of migration, (2) the intensity of migration, and (3) the contrast of the element.

GEOCHEMICAL CLASSIFICATION OF ELEMENTS ON THE BASIS OF THEIR SUPERGENE MIGRATION

A. Aerial Migrants

A_1. Active (forming chemical compounds)
   O, H, C, N, and I
A_2. Inactive (not forming chemical compounds)
   A, Ne, He, Kr, Xe, and Rn

B. Aqueous Migrants

B_1. Very mobile anions, with K_x between $10\,n$ and $100\,n$
   S, Cl, B, and Br
B_2. Mobile, with $K_x = n$
   B_{2a}. Cations: Ca, Na, Mg, Sr, and Ra
   B_{2b}. Anions: F

B_3. Weakly mobile, with $K_x = 0n$

    B_{3a}. Cations: K, Ba, Rb, Li, Be, Cs, and Tl

    B_{3b}. Mostly anions: Si, P, Sn, As, Ge, and Sb

B_4. Mobile and weakly mobile in oxidizing media ($K_x = n$ to $0.n$), and inert in strongly reducing media (K_x less than 0.1)

    B_{4a}. High mobility in acid and weakly acid oxidizing waters and low mobility in neutral and alkaline waters (predominantly cations): Zn, Ni, Cu, Pb, Cd, Hg, and Ag

    B_{4b}. Energetic migration in both alkaline and acid waters — more energetic in alkaline than in acid (predominantly anionic): V, U, Mo, Se, and Re

B_5. Mobile and weakly mobile in reducing, colloidal media ($K_x = n - 0.n$) and inert in oxidizing ($K_x = 0.0n$): Fe, Mn, and Co

B_6. Poorly mobile in most environments ($K_x = 0.n - 0.0n$)

    B_{6a}. Weak migration resulting in formation of chemical compounds: Al, Ti, Zr, Cr, rare earths*, Y, Ga, Cb, Th, Sc, Ta, W, In, Bi, and Te

    B_{6b}. Not forming or rarely forming chemical compounds: Os, Pd, Ru, Pt, Au, Rh, and Ir

This classification scheme is applicable to the supergene zone on the continents. Within each group, the elements are arranged in descending order of their abundance, which generally corresponds to their decreasing importance in supergene processes.

Only a few outstanding features are reflected in this classification. Because of the amphoteric nature of some elements, considerable departures are possible in some environments. For example, Co is placed in the B_5 group together with Fe and Mn, but because of certain peculiarities of its behavior in the supergene zone it could be placed together with Cu, Zn, and others in the group B_{4a}. Chromium, placed in the B_6 group of poorly mobile elements, acquires considerable mobility in deserts and is then analogous to the elements of the B_{4b} group.

Some poorly mobile elements like zirconium or yttrium may move in the form of organic complexes. The intensities of their migration increase in tundra swamps and similar environments. The migration of metals of the B_4, B_5, and B_6 groups intensifies in the sulfatic waters of the oxidation zone of sulfide or deposits.

Consequently, our classification scheme accounts only for the most common migration of elements in the supergene zone. Under

Table 27. Basic Geochemical Types of Epigenetic Processes

Aqueous migrants	Aerial migrants			
	O_2		CO_2, CH_4	H_2S
	Oxidizing		Reducing, without H_2S	Reducing, with H_2S
	In rocks containing reducers	In rocks not containing reducers		
H^+, SO_4^{2-}, Fe^{2+}, Zn^{2+}, Cu^{2+}	Sulfatic	–	–	Sulfatic, sulfidic
H^+, HCO_3^-, organic acids	Oxidation by acid waters	Acid	Carbonate-free colloidal	–
Ca^{2+}, HCO_3^-, Mg^{2+}	Oxidation by neutral weakly-mineralized waters	Neutral, calcareous	Carbonatic colloidal	–
Cl^-, SO_4^{2-}, Na^+	Oxidation by neutral strongly mineralized waters	Chlor-sulfatic	Oxysalt-bearing colloidal	Oxysalt–sulfide-bearing
Ca^{2+}, SO_4^{2-}	–	Gypsiferous	Oxysalt-bearing colloidal	
Na^+, HCO_3^-, OH^-, SiO_2	Oxidation by soda waters	Sodic	Sodic, colloidal	Soda and hydrogen sulfide

special conditions, the elements of the B_4, B_5, and B_6 groups may migrate more energetically.

CLASSIFICATION OF SUPERGENE PROCESSES

The same epigenetic processes may develop in different environments — in soils, in weathered crusts, or in water-bearing strata. A microbiologic reduction of sulfates (sulfidization), which generates hydrogen sulfide, takes place alike in marine oozes (Black Sea), in the muds of saline lakes, in marshes, or in underground aquifers. Salts accumulate in lakes and marshes. Soda is formed in soils and in deep-seated aquifers. Conversely, several distinct processes may develop side by side in the same natural medium.

A vertical differentiation is particularly significant. The processes operating in the upper strata of soils or crusts are drastically different from those operating in the lower strata.

Since epigenesis is common to all natural media, it is possible to group epigenetic processes into relatively few classes (Table 27). Each of these classes is common to all media — soils, weathered crusts, the terrigenous sediments, and the catagenetic subzone. Some occurrences of various processes are summarized in Table 28.

For the purposes of classification, each geochemical type of epigenetic process is characterized by a definite association of aeolean and aqueous migrants which move together in waters. The principles of this classification are self-explanatory (Table 27). Although the table lists only the basic types, the classification can be extended to cover more details. The following example illustrates this extension.

A special type of epigenetic process is encountered in coal regions as a result of the oxidation of coal. In some instances, the oxidation proceeds so energetically that the evolved heat starts a spontaneous combustion of coal. This produces a very peculiar epigenetic alteration of rocks. Since oxygen is deficient, coal forms various unburnt gases, including ammonia, sulfur dioxide, hydrogen sulfide, and various hydrocarbons. The surface of strata overlying underground coal fires becomes strongly heated. Various gaseous products escape upwards along the fractures. Various secondary minerals become deposited in fissures, such as sulfur and salammoniac. The oxidation of native sulfur brings about the formation of sulfuric acid and thus starts the sulfatic process. The rocks become intensely altered: whitened and coated with alums and other aluminum minerals.

The process fits easily in our classification scheme. It belongs to the type of such gases as NH_3, SO_3, CH_4, and H_2S. The SO_4^{2-}, Al^3 , and occasionally Fe^{3+} are aqueous migrants. Such conditions occur, for example, in Jurassic coal veins in the Yagnob river valley in Tadzhikistan.

The weathered crust developing over ultrabasic rocks is characterized by a specialized epigenetic alteration — the accumulation of silica, magnesium, and nickel in the lower levels of the crust.

Table 28. Occurrence of Basic Types of Epigenetic Processes

Vertical subzones	Epigenetic process					
	Sulfatic	Acidic	Neutral carbonatic	Chlor–sulfatic	Gypsiferous	Sodic
Soil	Over sulfide ore bodies	Meadow podzols; red, gray, brown soils of forested steppes; salt flats	In steppes: cher-nozems, chest-nut-brown, or gray soils	Upper layers of some salt marshes in steppes and deserts	Ancient salt mar-shes on terraces	Salt marshes
Weathered crust	Oxidation zone of sulfur and sul-fide ore depos-its; over pyritic shales and clays	Carbonate-free rocks in temper-rate and tropical humid climates	In steppes and deserts	Over salt-bearing rocks and salt de-posits	Gypsum cappings over salt bodies	—
Continental de-posits	—	Deluvium in hu-mid climates	Deluvium and proluvium in arid climates	Saline sediments in deserts		Red sediments
Subzone of cata-genesis (aquifers)	Sulfide-bearing rocks, flushed by oxygenated waters	Carbonate-free rocks in humid climates	Carbonate-rich rocks (lime-stones, marls, loesses)	Salt-bearing rocks; aquifers in deserts	Places where deep chlor–calcic waters mix with surficial sulfatic waters	Wastes of steppes; deep-lying sands with artesian waters

Table 28 (continued)

Vertical subzones	Epigenetic processes						
	Carbonate-free colloidal	Carbonate-rich colloidal	Saline colloidal	Gypsiferous colloidal	Sodic colloidal	Oxysalt-sulfidic	Sodic with H_2S
Soil	In swamps of taiga, tundra, or tropics	Meadow and swamp soils of northern steppes; carbonate-rich on meadows and in swamps of forest and tundra zones	Salt marshes with weakly-reducing quality	Gypsum horizons of meadows	Soda-salt marshes on meadows	Lower horizons of salt marshes	Salt marshes.
Weathered crust	Plains of northern taiga and tundra	—	—	—	—	—	—
Continental deposits	Alluvium in humid climates	Alluvium in forested and northern steppes	Muds in saline lakes with weak reducing media	—	Muds of soda lakes	Muds of saline lakes with strong reducing media	—
Subzone of catagenesis (aquifers)	Carbonate-free rocks in predominantly humid climates	Carbonate-rich and sulfate-poor rocks	Rocks in arid climates	Aquifers in gypsum strata	—	Deep oil deposits	Oil deposits, bituminous limestones, and shales in process of destruction; flushed by weakly mineralized waters

The following minerals form there: Opal, $SiO_2 \cdot nH_2O$; magnesite, $MgCO_3$; sepiolite, $Mg_3(SI_4O_{10}) \cdot H_2O \cdot nH_2O$; hydromagnesite, $Mg_5 \cdot (CO_3)_4(OH)_2 \cdot 4H_2O$; kerolite (white serpentine), $Mg_4(Si_4O_{10}) \cdot (OH)_4 \cdot 4H_2O$; and garnierite, $Ni_4(Si_4O_{10})(OH)_4 \cdot 4H_2O$.

Ginzburg (1957) studied these phenomena in detail on serpentinites of the Urals. Both the colloidal and the ionic reactions were found to be taking place side by side in the same rock. To explain this, let us consider the case of an aquifer which contains an insufficient amount of hydrogen sulfide to convert all the available ferric iron to a sulfide. Some of the reduced iron will thus migrate in the form of iron bicarbonate. In other words, the fixation of iron as pyrite is simultaneous with its migration when hydrogen sulfide is deficient.

Reducing media entirely lacking hydrogen sulfide are not uncommon, but wherever organic remnants have been buried the reducing media are rich in hydrogen sulfide. Thus, the same water-bearing level may accommodate both processes, the sulfide-ionic and the colloidal. In fact, there exist transitions between the two.

Chapter 10

Epigenetic Processes of Oxidation

THE SULFATIC PROCESS

The sulfatic process develops in rocks containing either sulfides or free sulfur. Attacked by oxygenated ground waters, these oxidize and produce sulfuric acid:

$$FeS_2 + 7O + H_2O \rightarrow FeSO_4 + H_2SO_4,$$
$$2H_2O + 2S + 3O_2 \rightarrow 2H_2SO_4.$$

As a result, waters become strongly acid, their pH decreasing to 2 and even to 1 and their sulfate ion concentration rising. Aluminum (which remains inert in most other processes), iron, copper, zinc, and other metals acquire mobility as the pH drops. Other possible oxidizers include free sulfuric acid and ferric and cupric sulfates. The Eh varies all the way up to $+0.7$ V.

The Eh–pH relationships for waters in the oxidation profile of ore deposits are shown in Fig. 35. An important role in the sulfatic process is played by sulfur bacteria which have been identified in the acid waters of ore deposits and coal seams. Due to their activity, the pH may drop to 1. They best thrive at a pH between 3 and 4, but can survive at as low a pH as 0.2 (Birshtekher, 1957).

Ferrous sulfate is one of the end products of the oxidation of pyrite and chalcopyrite, since it does not oxidize any further when in a strongly acidic medium. However, it oxidizes to ferric sulfate, $Fe_2(SO_4)_3$ under the action of sulfur bacteria:

$$2MeS + 2Fe_2(SO_4)_3 + 2H_2O + 3O_2 \rightarrow 2MeSO_4 + 4FeSO_4 + 2H_2SO_4.$$

According to Becking et al. (1963), an inorganic oxidation of pyrite lowers the pH of the solution to 3.1 at 650 mV. In the

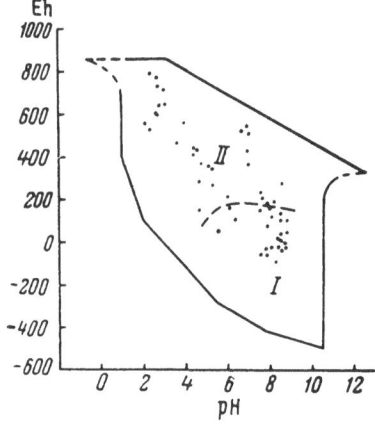

Fig. 35. Eh–pH parameters of mine waters. An attempt to differentiate between waters of primary ores (I) and those draining from oxidation zones (II) (after Becking et al., 1963).

presence of bacteria the pH drops to 2, while the Eh rises to 850 mV.

Some experiments suggest the feasibility of microbiologic oxidation of other sulfides also, including molybdenite, chalcopyrite, chalcocite, covellite, and bornite.

According to microbiologists (Kruznetsov et al., 1963), pyrite and chalcopyrite oxidize faster in the presence of bacteria. The first stage of oxidation to $FeSO_4$ may well be a strictly chemical reaction, but the oxidation from $FeSO_4$ to $Fe_2(SO_4)_3$ is biochemical.

The sulfatic process develops equally well in soils, weathered crusts, and catagenetic subzones. The oxidation zone of sulfidic ore deposits is in reality a variation of the weathered crust.

The oxidation zone of sulfidic ore deposits. The oxidation of ores containing sulfides of copper, iron, arsenic, antimony, and other chalcophite elements intensifies the acidity of the reaction and the formation of typical mineral assemblages such as sulfates, phosphates, and hydroxides. Much has been written on the subject, since it has a very definite practical value. The monograph by Smirnov (1955) deals with this question. The geochemistry of the oxidation zone has been reviewed by Shcherbina (1955). Conclusions reached in such investigations are too numerous to be discussed here. It can only be stressed that the structure of the oxidation zone is influenced by a multitude of factors, including the climate, the composition of primary ores, the texture, and the composition of the host rocks.

Typically, the oxidation zones develop in stages, but the tendency is always in the direction of rising pH. The mineral composition of the oxidizing ores changes accordingly. The transition from acidic to neutral to alkaline media was expressed by Shcherbina in the form of the following mineral series:

For iron

Pyrite	Melanterite	Coquimbite	Jarosite	Limonite
$FeS_2 \longrightarrow$	$FeSO_4 \cdot 7H_2O \longrightarrow$	$Fe_2(SO_4)_3 \cdot 9H_2O \longrightarrow$	$KFe_3(OH)_6(SO_4)_2 \longrightarrow$	$2Fe_2O_3 \cdot 3H_2O$

For lead

Galena	Anglesite	Cerussite	Pyromorphite	Vanadinite
$PbS \longrightarrow$	$PbSO_4 \longrightarrow$	$PbCO_3 \longrightarrow$	$Pb_5(PO_4)_3Cl \longrightarrow$	$Pb_5(VO_4)_3Cl$

For copper

Chalcopyrite	Bornite	Covellite	Chalcocite
$CuFeS_2 \longrightarrow$	$CuFeS_4 \longrightarrow$	$CuS \longrightarrow$	Cu_2S
Copper sulfides	Chalcanthite	Ellite	Malachite
$(CuS, Cu_2S) \longrightarrow$	$CuSO_4 \cdot 5H_2O \longrightarrow$	$Cu_5(OH)_4P_2O_8 \longrightarrow$	$CuCO_3Cu(OH)_2 \longrightarrow$
	Chrysocolla	Uzbekite	
	$CuSiO_3 \cdot 2H_2O \longrightarrow$	$Cu_3V_2O_8 \cdot 4H_2O$	

For zinc

Sphalerite	Goslarite	Smithsonite	Calamine
$ZnS \longrightarrow$	$ZnSO_4 \cdot 7H_2O \longrightarrow$	$ZnCO_3 \longrightarrow$	$Zn_2(OH)_2SiO$

Vertically, the oxidation zone is never homogenous, but consists of several subzones: the surficial subzone, the subzone of leaching, the subzone of rich oxidized ores, etc. (Smirnov, 1955). The system of subzones is never the same for all ore deposits.

The development of an oxidation zone is a lengthy process which may last several geological periods. Different processes succeed each other as the climate and tectonism change. Upper portions are eroded away and the descending oxidation imposes new aspects on the already oxidized lower portions. All this points to the need of a historical approach to this complex problem.

The great age of the oxidation zone has been proved for many ore deposits in the southern Urals, central Kazakhstan, Karamazar, and northern Siberia. In some ore deposits of Karamazar, oxidation zones started to develop in the Upper Cretaceous period.

Compared to the other waters of a given region, waters draining from the oxidation zone display characteristically lower pH values and higher contents of SO_3^{2-}, Zn^{2+}, Cu^{2+}, MoO_4^{2-}, etc. Aqueous haloes of zinc or copper dispersions stretch for hundreds of meters and up to 2 km. Regions can be prospected for heavy metals by determining the content of these metals in waters.

Investigations by Udodov et al. (1962) have shown that the heavy metal content of waters from oxidation zones is 10 times greater than that of the "background." For example, a heavy metal content of 750-850 mg/liter was found in the oxidized zone of the Zmeinogorsk district, as compared to a "background" content of only 70-170 mg/liter. In the Norilsk district, similar analyses gave 200 mg/liter SO_4^{2-} and 10 mg/liter, respectively.

The metal content of halo waters fluctuates seasonally and also with the weather. In the Altai and Sayan Mountains it increases after rains. This fact should be taken into account during prospecting.

According to Shcherbakov (1959), the oxidation of calcareous quartzites of the Krivorozh'e imparts carbon dioxide to waters:

$$CaCO_3 + H_2SO_4 \rightarrow CaSO_4 + H_2O + CO_2 \uparrow.$$

In limestones, this is a common oxidation reaction of the sulfides of the veins. Elsewhere, it accounts for the karst topographies and cavities (ore karst) which are so typical of Tyan-Shan, Talass Ala-tau, and Karamazar.

In permafrost regions, the exothermal reactions of oxidation promote the thawing of the ground. Most of the ore deposits in the Norilsk region lie in cavities of the thawed-out rock. The data on file at the Permafrost Institute indicate that the redox potential fluctuates between 0.5 and 1.5 V in the frozen ground. Udodov et al. (1962) stated that the mean annual temperature of the upper portions of the Ek-Khai sulfide ore deposit was −6°C, while that of the surrounding rocks was −10°C.

The evolution of heat during the oxidation of sulfides sometimes results in spontaneous combustion (in Spain or in America). The weathered crust on pyritic shales, clays, and coals is characterized by strongly acid waters. Enriched in aluminum and ferric iron, these waters deposit incrustations of: alum, $KAl(SO_4)_2 \cdot 12H_2O$; jarosite, $KFe(SO_4)_2(OH)_6$; alunogen, $Al_2(SO_4)_3 \cdot 16H_2O$; rhomboclase, $FeH(SO_4)_2 \cdot 4H_2O$; leucoglaucite, $FeH(SO_4)_2 \cdot 2H_2O$; coquimbite, $Fe_2(SO_4)_3 \cdot 9H_2O$; etc.

According to Goleva (1959), similar incrustations occur in the black Maikop clays of the northern Caucasus and in the Oligocene melilite slates of the Carpathian Mountains. She stated that the pH dropped from 6.9 to 2.9 as the percolating waters became enriched in Al (up to 40 mg/liter), Fe (up to 50 mg/liter), As, Pb, and Mn.

When coal seams are opened up by mine workings, oxygen pene-

trates the strata and the mine waters become oxygenated. Disseminated pyrite oxidizes and waters turn acid; their pH drops to below 3,
and their iron content rises and may reach 10 mg/liter. Iron bacteria play an important role in the oxidation of pyrite (Birshtekher,
1957).

When attacked by percolating atmospheric waters, sulfur oxidizes and produces sulfuric acid, which, in turn, attacks the rocks.
Iron and aluminum become mobile. Iron cappings develop. Secondary minerals of oxidation include jarosite, alunite, and loweite.

The action of sulfuric acid on limestones produces gypsum.
Some of the largest gypsum deposits in the world, like those of
southwestern Iran, were formed in this way.

Sulfatic processes in the catagenetic subzones
develop only in the strata permeated by oxygenated waters. Because of such interaction, black to gray pyritic sandstones are
bleached. Their pyrite crystals are replaced with ferric hydroxide.
The color of oxidizing rock becomes a mottled greenish brown.
Similar color changes take place in argillaceous rocks and alevrolites. The development of this process is intensified during tectonic
uplifts when the circulation of ground waters is speeded up. The pyritic Mesozoic and Cenozoic rocks of Central Asia have widely oxidized in regions of alpine uplifts. Boundaries between the oxidized
and the reduced rocks can easily be traced by their colors (Fig. 36).

A good example of sulfatic oxidation was cited by Koval'chuk and
Sokolova: On the eastern slopes of the Urals, the oxygenated
ground waters oxidize lignitic clays which contain disseminated
pyrite and marcasite. As a result, acid sulfatic waters form and
then become colloidal. The oxidation progresses slower in the

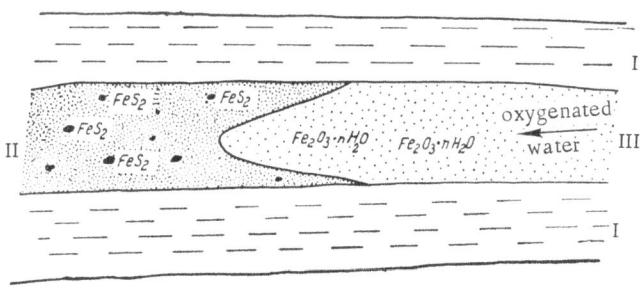

Fig. 36. Interstratal oxidation of sandstones and clastics. (I) Clays. (II) Pyritic sandstone. (III) Oxidized, limonitic sandstone.

buried strata than in the weathered crust. Therefore, the reactions going on in the buried strata are not quite as acid. As fast as it forms, sulfuric acid is neutralized in the course of the cationic exchange with the enclosing rocks. Thus, the sulfatic process seldom extends beyond the immediate vicinity of the oxidizing veins. Generally speaking, a neutral oxidation is typical in water-bearing strata.

Gypsum is a characteristic epigenetic mineral. Its plates fill the fissures in oxidizing clays and alevrolites. The Paleogene strata of the Fergana Basin and the green Turonian clays of Kyzylkum are examples. It may be mentioned that the green clays shade into gray below the oxidation zone.

S u l f a t i c   p r o c e s s e s   i n   s o i l s have not been investigated as thoroughly as the processes in the catagenetic subzone. It was established, however, that pyrite oxidizes as the pH drops. Soils acquire rusty, red-brown colors. Such soils, when developed over sulfidic ore bodies, should be studied carefully. The soils of Australian marshes measured pH 2 at 860 mV.

T h e   p a r t i a l   o x i d a t i o n   o f   p y r i t e. As already stated, ferrous iron is stable in strongly acid media. For this reason, the oxidation of pyrite is rarely complete. Sulfur oxidizes before iron does. A partial oxidation is common in the supergene zone. It has been best studied in the oxidation zones of sulfuric ore deposits where iron migrates in form of ferrous sulfate, $FeSO_4$.

Recently, additional data became available on oxidation in other environments. For example, at Kyzylkum we observed the oxidation of gray pyritic shales of Lower Turonian age, which are exposed in a quarry. The clays became fractured on drying and shrinking throughout the entire exposed thickness of 10 m. Films of limonite developed along fractures and in spots, the latter up

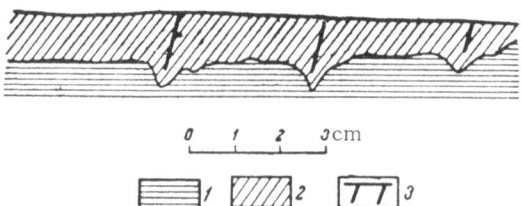

Fig. 37. A partial and a complete oxidation of Lower Turonian clays. (1) Gray clays; (2) green, completely oxidized clays; (3) red ferruginous films.

to several meters in diameter. The mechanism of this superficial iron enrichment is revealed when a lump of semioxidized clay is crushed open. A core of unaltered gray clay is surrounded by a zone of green oxidation which is about 1 cm thick. The green alteration extends into the core, making inroads along the fractures which are lined with limonite (Fig. 37).

Obviously, the alteration set in after the exposure had been made. Drying of the clay resulted in shrinking and fracturing. Capillary waters held by the clay became oxidized by the air pene- trating into the fractures. This brought about the oxidation of the pyrite. Since the clays contained no calcium carbonate, the reac- tion could be acid. Dissolved ferrous iron diffused upwards and there became oxidized to ferric. Limonite precipitated.

In this example, different geochemical media succeed each other in a space of 1 cm and comprise an "epigenetic microzoning." An analogous zoning, measurable in hundreds and thousands of meters, consists of gray pyritic rock enclosed on top by partly oxidized rocks of mottled colors and then by a limonitic zone. This last zone coincides with the region of infiltration of oxy- genated waters. Uraniferous mineralization is commonly related to this type of zoning, as will be discussed in Chapter 13.

The criteria for the recognition of sulfatic processes are:

1. Typical mineral assemblages: sulfates, carbonates, and phosphates of heavy metals; basic sulfates of iron and aluminum; iron cappings over the ore deposits
2. Strongly acid media with pH below 4
3. Abundance of the sulfate ion and scarcity of chloride ion in waters; considerable contents of heavy metals
4. Rusty, brown color of rocks is indicative but not conclusive

ACID PROCESSES

These processes develop in moderately acid waters, which are buffered with carbonic or humic acids or other similar substances. Their pH varies between 4 and 6.5, while the Eh values are between +0.4 and 0.7 V. Weathering is the most important result of such processes. During weathering, cations of the rock are replaced with the hydrogen ions. Calcium and sodium are the first to be leached in acid media. They are followed by Mg, K, etc. At the

same time, the existing minerals hydrate. Rocks containing bivalent iron or some other reducers turn red to brown in color on the
oxidation of such components. Feldspars and other aluminosilicates
are transformed into clays. This a typical reaction.

The weathering is qualified as kaolinitic, bentonitic, etc.,
depending on the nature of the clay minerals formed. In deep-
seated environments, silica may be removed and the rock residually enriched in compounds of iron, aluminum, and other less
mobile elements. The migration of finely divided clay minerals
in the suspension and the formation of plasma are characteristic.

In regions with humid and temperate-to-hot climates, acid
processes develop commonly in soils, weathered crusts, and
water-bearing strata provided that the rocks contain no sulfides,
gypsum, carbonates of calcium or magnesium, or water-soluble
salts. The humidity of the climate hastens the leaching of the water-
soluble components. The relatively high temperatures favor the
development of lush vegetation and a rapid decomposition of dead
organic matter. Waters become enriched in carbon dioxide and
other products of organic decomposition.

At the present time, acid processes are best developed in the
weathered crusts in the humid tropics. There the decomposition
of the remnants of lush vegetation produces large quantities of
carbon dioxide and humates. Heavy atmospheric precipitation —
some 2000 to 10,000 mm annually — coupled with high temperatures
decomposes rocks for several hundred meters. The resulting
crust is usually leached free of all monovalent and bivalent elements
and silica. The hydroxides of iron and aluminum tend to accumulate.

In regions with hot climates, waters carry relatively light loads
which consist mainly of silica. Typomorphic elements here are the
hydrogen ion and the silica. Characteristically, silica accumulates in the form of opal and chalcedony concretions. Iron and
manganese precipitate in the form of hydroxides. In some instances,
aluminum accumulates in the form of hydroxides, kaolin, or hal-
loysites.

In the USSR, all this can happen only in the Batum region. In
the geologic past, this type of weathering was widespread. Ancient
crusts occur in the Urals, Kazakhstan, the Far East, and the
European part of the USSR. These crusts have been developing in
different periods since the Precambrian (Ginzburg, 1949).

In tropical climates, ore mineralization is frequently related to the acid weathering. Eluvial bauxites, consisting of aluminum and iron hydroxides, comprise the upper part of the acid crust developed in the nepheline syenites. Kaolinitization progresses intensely in the lower parts of the same crust. Examples are found in the USA (Arkansas) and in New Guinea. Rare earths occasionally accumulate with aluminum. Kaolin deposits form during the weathering of granites.

During the alteration of ultrabasic rocks, iron and aluminum accumulate in the upper part of the crust and nickel in the lower. The iron deposits of Cuba and the nickel deposits of New Caledonia are examples. In the USSR, similar nickel deposits are developed in the weathered serpentinites of the Urals (Ginzburg, 1949).

The humid and predominantly hot tropical climates of the past geologic periods can be identified by acid crusts. Such crusts corresponded to flat topographies, quiescent tectonism, and low mineralization of ground waters.

In humid and temperate climates, acid migration is slowed down. Weathering penetrates to shallower depths; the leaching of cations is considerably less. The waters are somewhat more mineralized and their pH is close to 7. Due to the formation of limonite, the color of weathered rocks is generally brown. Such are the weathered moraines of central USSR and the igneous rock eluvia of the Urals.

The migrational series of Table 26 (p. 151) were worked out for the weathering conditions in humid and temperature climates. The extent of leaching is illustrated by deposits of quartz sand, which run 95 to 98% pure silica.

Acid migration is common in all soils in all climates. The fact that waters enriched in humates leach various elements should be taken into account during soil sampling. Only the deeper soil levels lying at a depth of 0.5 to 1.0 m below the surface should be sampled.

Acid leaching of terrigenous sediments tends to produce colloids. For this reason, this process will be considered elsewhere.

In the catagenetic subzone, the effect of oxygenated waters in regions of active water circulation is noticeable down to depths of 100 m or more. However, this effect is subdued in comparison with the same effects in soils or in weathered crusts.

In general, the trends of alteration by acid processes remain very much the same as those due to the sulfatic processes, and result in the removal of cations and limonitization.

The criteria for the recognition of acid migration are:

1. Intense rock alteration, which nevertheless preserves some relict structures
2. Red, brown, and mottled colors
3. Absence of carbonates
4. Epigenetic assemblages, consisting of iron hydroxides, clay minerals, opal, chalcedony, secondary quartz, and, rarely, aluminum hydroxides

Depending on the composition of the rocks, two subgroups are distinguished: (1) acid oxidation of gray and green rocks which contain organic matter, bivalent iron, and other reducers; and (2) acid weathering of rocks not containing any reducers.

THE NEUTRAL CARBONATIC PROCESS

This process involves the migration of oxygenated waters which are quite mineralized. Calcium carbonate is their predominant component. The properties of such waters are oxidizing. The uranium, strontium, sodium, magnesium, and sulfate-ion migrate readily, while the iron, aluminum, and humates migrate more sluggishly. The typomorphic ions here are calcium, magnesium, and bicarbonate.

Characteristically, this process develops in the soils, in the weathered crusts, and in the terrigenous sediments provided that the climates are moderately dry. In the alpine regions of Central Asia, Kazakhstan, Transcaucasia, or Transbaikalia, it frequently develops in loesses or in weathered crusts.

The same epigenesis is found occasionally in other climates. In humid climates, it develops in the vicinity of limestones. In deserts, it is found in the vicinity of volcanic rocks which were deeply dissected by erosion.

The carbonatic weathered crust develops commonly in arid regions because of the scarcity of water. Calcite, formed during the decomposition of rocks, is not carried away but accumulates in place. It imparts to the weathered crust its characteristic light color and calcareous character.

Many loesses and loessoid rocks* are good examples of this

*Obviously, not all the loesses constitute a weathered crust. The loessoid terrigenous sediments may be the products of the loess erosion.

type of crust. In Belorussia the crust, developed over the lacustrine deposits since the Early Jurassic period, is 13 m thick in places. Its mineral composition is the hydromica–beidellitic; locally it contains mollusk shells and ironstone concretions. According to Samodurov (1963), the cover of loess and loessoids is typical of the Quaternary glaciation on the Russian Plain and in the foothills of the eastern Carpathians.

Solution cavities are created in water-soluble rocks. Calcium carbonate, leached out in one place, is redeposited in another.

In the USSR, karst topographies are found in many regions and belong to all ages, according to Rodionov. The development of karst depends on the climate and the tectonics. It develops most rapidly in temperate-to-warm, humid climates. It develops most sluggishly in polar and arid regions. It is stopped by subsidences. Several "levels" of karst develop in the course of repeated uplifts and subsidences. Different tectonic structures impose their own characteristics on its development. The erosional karst is predominant on the plains. In mountainous regions, caverns are formed on fissures in the subsurface. An argillaceous residue left behind on the leaching of calcium carbonate is reddish brown in color.

In steppes, the interstitial evaporation of water carrying calcium bicarbonate leads to calcareous impregnation in capillary fringe zones. Such calcareous impregnations are characteristic of the Central Russian uplift. We have found such layers just south of the city of Orel: a lime cemented stratum lying under 3 m of soil and clayey subsoil. On the plains of Central Asia, such processes operated during the Neogene and the cold Pleistocene epoch (Perel'man, 1959). According to Sidorenko (1958), similar processes were active in the deserts of Mexico.

These impregnations have received various names — the "calcareous incrustation," the "paree" (in the Near East), and the "caliche" (in America). Sidorenko (1959) observed such impregnations in the Libyan desert. It is possible that these impregnations are formed by the ascending waters which dissolved the calcium carbonate in depths and brought it up in the form of calcium bicarbonate. As a result of this carbonatic epigenesis, stringers of strontianite, $SrCO_3$, fill the fissures in limestones. Among other epigenetic formations we might mention the clay-derived calcite which fills fissures and veins and also forms concretions.

Because the air and water entrapped in soils contain much

carbon dioxide (derived from oxidizing vegetal remnants), calcium carbonate is being removed from the upper horizons of chernozems and chestnut-brown soils. It is transported in the form of calcium bicarbonate. Having arrived at the lower strata, where the content of carbon dioxide is much lower, calcium bicarbonate breaks down and precipitates the calcium carbonate. This is the mechanism of development of calcareous beds lying from 0.5 to 10 m below the surface.

Studying the calcareous impregnations in loesses and loessoids of the Central Russian forested steppe, Dobrovol'skii (1957) identified three types of impregnation: (1) disseminations; (2) spotty; and (3) segregations. Many segregations display metacolloidal structures which suggest the colloidal precipitation of calcium carbonate. Spectrographic analyses invariably disclose the presence of strontium, barium, copper, manganese, zirconium, chromium, and, occasionally, gallium and nickel. Copper, strontium, and manganese probably migrate together with calcium carbonate.

According to Gvozdetskii (1954), this type of epigenesis was common in all parts of the USSR during all geologic epochs. The criteria for the recognition of such processes are: (1) karst cavities; (2) red residual clays in solution cavities of the limestones; and (3) secondary calcium carbonate concretions, tuffs, and stringers. On the other hand, any evidence of the aqueous transfer of aluminum, iron, or humic substances precludes the development of the neutral carbonatic processes.

Some commercial deposits may be produced in the course of such processes. Phosphatic limestones and marls are particularly interesting in this respect. Descending surface waters leach out calcium carbonate and leave calcium phosphates in cavities together with the residual clays. This is the origin of many phosphorites. Because waters carrying calcium bicarbonate also carry molybdenum, uranium, and fluorine, secondary dispersion halves form around ore deposits. This must be kept in mind during geochemical exploration.

The process may be divided into two types: (1) the neutral carbonatic process in rocks containing some active reducers which become oxidized; during this process the rocks redden in color and calcium carbonate is removed; and (2) the same process, but in already oxidized rocks involves only the redistribution of calcium carbonate.

The second type develops in soils and aquifers if the conditions are weakly oxidizing. Manganese becomes reduced to the bivalent state and then migrates in the form of manganese bicarbonate, $Mn(HCO_3)_2$. Eventually, manganese reprecipitates in the form of black oxides and hydroxides. Travertines, which may be colored black* by such impurities, accumulate at the points of the issuance of springs. Calcareous tuffs, like those of the Tomsk region, may contain up to 1% Mn and Sr, 0.03% Co, 0.01% Ni, Cr, Ba, V, Ti, and traces of S, Pb, Cu (Udodov et al., 1962).

The ground waters of the city of Khmelnic described by Babinets (1961) should also be weakly oxidizing, since their manganese content varies between 3.5 and 14 mg/liter. The "background" manganese content in waters of the region is only about 0.5 mg/liter.

Since iron remains immobile during this process, the rocks retain their reddish-brown colors.

Smudges of black manganese hydroxides, common on the outcrops of red rocks, indicate how widespread these weakly oxidizing conditions were and how easily manganese migrated in the epochs of the red rock formation. The epigenetic calcite, containing up to 18.5% of the admixed rhodochrosite, replaces crushed grains of quartz, feldspars, biotite, and clay minerals in sandstone of the Usha series† (Kopeliovich, 1962).

Sedimentary carbonates and the bicarbonatic ground waters were the probable sources of the epigenetic calcite. A number of other mobile elements could have been leached together with calcite.

THE CHLOR-SULFATIC PROCESS

The process involves strongly mineralized waters which carry chemical loads of several grams to several hundred grams per liter. Their composition varies from predominantly sulfatic with a minor amount of chlorides to predominantly chloridic with a minor amount of sulfates. Their reaction is neutral and, as a rule, these waters are oxygenated. Chlorine, sulfate sulfur, boron, sodium, strontium, and, locally, uranium, chromium, and iodine migrate easily in such waters. Iron, copper, aluminum, and titanium are quite immobile.

Chlor-sulfatic epigenesis accounts for the deposition of large

---

*As a rule travertines carry very little iron, if any.

†The Usha series of the Dnepr region is a transition between the Riphaean (U. Proterozoic) and Cambrian periods.

Fig. 38.   Karst sinkholes in Neogene salt-impregnated alevrolites, Kyzylkum (after Schafeeva).

masses of industrial salts* in regions of positive relief and active water circulation. In the Jurassic period and again in the Tertiary period, this process produced such Central Asiatic salt deposits as those of Ak-Chop and Ak-Bel' in Fergana or of Hodzha-Mumyn Mountain in southern Tadzkikistan.

Ground or surface waters flowing in or over salt measures are actually brines containing up to several hundred grams per liter. Therefore, the points of the issuance of such waters are incrustated with salts. As it were, streams flow over salt bottoms and between the salt banks.

Salt karst develops as the result of salt removal from the rocks. Since the water not only erodes but dissolves as well, the surface develops sharply serrated structures.

The above-mentioned Hodzha-Mumyn Mountain is a typical example of such a salt karst. The mountain rises 900 m above the surroundings and 1400 m above sea level. The highest salt mountain in the Soviet Union, it is composed of salt- and gypsum-bearing strata of the Upper Jurassic period. In deserts, the salt karst may develop even in salt-bearing alevrolites (Fig. 38).

Since the intercalation of rock salt with gypsum or anhydrite is

_____

*Industrial salts include gypsum and anhydrite, rocksalt, potash-magnesium salts, sodium sulfate, and various borates. Salt measures include salt-impregnated strata as well as lenses of high purity.

typical of salt measures of all geologic epochs, it is only logical to expect that gypsum enriches residually when the rock salt is leached out. A gypsum capping of the rock salt is the result, which involves the hydration of anhydrite to gypsum.

At the present time, the chlor-sulfatic process is widespread in deserts and semideserts, producing salt lakes and salt-contaminated soils, although the oxysalt-sulfidic process is more characteristic for such a region (see below).

Concentrations rise as waters evaporate and various salts begin to fall out of the solution (Fig. 39). The waters first become chlor-sulfatic, then sulfato-chloridic. Their chemical loads may go as high as 200 g/liter. Precipitating out of the ground waters, the salts accumulate in the soils and subsoils where zones form. Calcium carbonates and gypsum precipitate near the ground water table; water-soluble salts precipitate closer to the earth surface. Should the depth to the ground water table be 1 m or less, gypsum and water-soluble salts would precipitate simultaneously.

The precipitation by evaporation is characteristic of many chemical elements — both the principal and the accessory ones. Accessory elements, concentrating in salt lakes and marshes, may form independent minerals. A considerable factor in the localization of salts in lakes and soils is the existence of salt accumulations in the vicinity.

The differences in the solubilities of salts are reflected by their irregular distribution in waters and soils (Fig. 40). Climate exerts a considerable influence. In general, the more humid the climate, the less salt contamination is to be found in soils and waters.

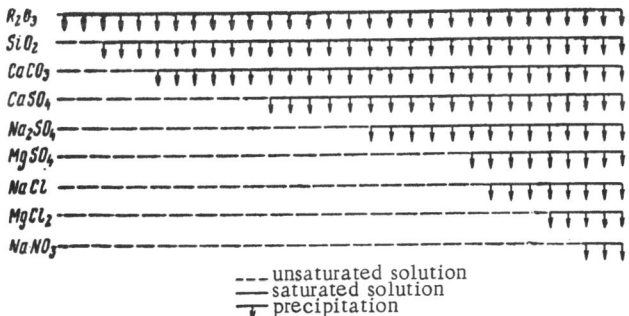

Fig. 39. A scheme of consecutive precipitation from evaporating ground waters (after Kowle, 1946).

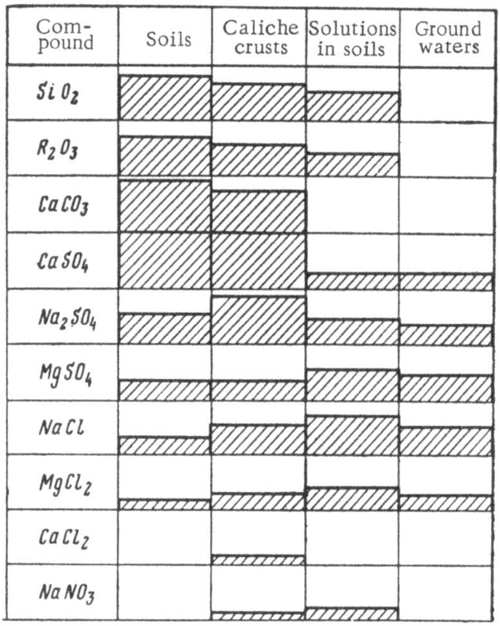

Compound	Soils	Caliche crusts	Solutions in soils	Ground waters
SiO_2				
R_2O_3				
$CaCO_3$				
$CaSO_4$				
Na_2SO_4				
$MgSO_4$				
$NaCl$				
$MgCl_2$				
$CaCl_2$				
$NaNO_3$				

Fig. 40. A scheme of differentiation of compounds precipitating during the evaporation of ascending ground waters (after Kowle, 1946).

The geochemistry of salt contamination in soils is controlled by various geologic factors. The process of concentration by evaporation is common on alluvial plains where the waters are supplied by surface streams. Such are the salt contaminations at Karakum, Zeravshan, Amu Darya delta, etc. In deserts, salting is largely restricted to faults, along which the deep waters ascend to become soil waters. Many marsh basins are depressions produced by faulting. Such marshes, of course, are different from the marshes fed by thermal or mineral springs. These often carry some rare elements.

Typical minerals deposited by the chlor-sulfatic waters are gypsum, celestite, and water-soluble salts. Precipitating, gypsum forms concretions and impregnations. In soils, its content may exceed 70%. The water-soluble salts are represented by: halite, $NaCl$; thenardite, Na_2SO_4; glaserite, $K_3Na(SO_4)_2$; langbeinite, $K_2Mg_2(SO_4)_3$; glauberite, $Na_2Ca(SO_4)_2$; astrakhanite, $Na_2Mg(SO_4)_2 \cdot 4H_2O$; kieserite, $MgSO_4 \cdot H_2O$; and vanthoffite, $Na_6Mg(SO_4)_4$.

There are quite a few variations of the chlor-sulfatic process, each of which is portrayed by a specific mineral assemblage. For instance, the following minerals form in borate deposits: kurna-kovite, $Mg_2B_6O_{11} \cdot 13H_2O$; asharite, $MgHBO_3$; boracite, $Mg_6B_{14}O_{26}Cl_2$; boronatrochalcite, $NaCaB_5O_9 \cdot 8H_2O$; inderborite, $MgCaB_6O_{11} \cdot 11H_2O$; etc.

The gypsum capping of the borate deposits usually contains asharite. Manifestations of the chlor-sulfatic process in the coastal deserts of Chile — the Atacama and Tarapaca deserts — are very interesting in view of the fact that these contain iodine and chromium: tarapakaite, K_2CrO_4; lautarite, $Ca(IO_3)_2$; dietzeite, $Ca_2(IO_3)_2(CrO_4)$; etc.

Epigenetic processes resulting in the precipitation of mirabilite, $Na_2SO_4 \cdot 10H_2O$, should be separated into a special group. The weathered crust of Kirghizia by Shcherbina, who found these to be common saddles between the mountains in the Tyan-Shan range. He came to a conclusion that glauberite, $Na_2SO_4 \cdot CaSO_4$, was formed directly from sodium sulfate waters in the stagnant lakes of the Neogene time. He found glauberite only below the weathered material, the latter containing mirabilite with relict glauberite. He holds that glauberite decomposes during weathering into gypsum and mirabilite.

De-dolomitization is another variation of the chlor-sulfatic process. It predominates in dolostones intercalated with gypsum. Waters carrying calcium sulfate react with dolomite:

$$CaSO_4 + MgCO_3 \cdot CaCO_3 = 2CaCO_3 + MgSO_4.$$

In this reaction, dolomite is destroyed and secondary calcite is deposited, while magnesium sulfate goes into solution. De-dolomitization is a common phenomenon in Central Asia, where it has been studied by Tatarskii (1953) and many others. For example, such phenomena were observed in the dolomites of the Bukhara level (Paleogene) in the Tadzhik and Bukhara-Khiva depressions. Secondary calcite and the solution cavities are the criteria for the recognition of the process, except where the dolostones are pyritic. The oxidation of pyrite protects the dolostones. An occurrence of magnesian lakes or marshes in the region is another indication. De-dolomitization, in turn, is indicative of arid climates, since calcium carbonate is subject to solution and removal in humid climates.

Another example of the process is the removal of salts from salt-impregnated clays along their contacts with water-bearing strata. According to Priklonsky and Oknina (1960), chlorides were removed completely and gypsum partially within two to four months.

Like the carbonatic, this chlor-sulfatic process may be divided into two types depending on whether the rocks are or are not already oxidized. A weakly oxidizing process, noted for the migration and concentration of manganese, is representative of the second type. At the present time, such processes are going on in the Neogene strata of Balkhyz in Turkmenia. There, in the Kyzyl-Dzhar ravine (northeast of the Er-Oilan-Duz salt lake) black bands of manganese hydroxides mark the places of the issuance of strongly mineralized waters in the red rocks. Black smudges are also common to all ancient red alluvials occurring in the western Uzbekistan and Turkmenia. This indicates the weakly oxidizing conditions which prevailed during the Neogene period on the flood plains and in deltas. These conditions favored the migration of manganese but not of iron.

The criteria for the recognition of the chlor-sulfatic process are the fissure fillings and impregnations of gypsum and water-soluble salts. These criteria may be used for the reconstruction of the paleogeographic conditions, since they imply the one-time existence of strongly mineralized waters and the possibility of an arid, hot climate.

In geochemical prospecting, the effects of chlor-sulfatic epigenesis should be carefully understood in order to avoid tracing "false anomalies." For example, the average uranium content in ground waters in deserts is $n \cdot 10^{-5}$ g/liter and in the mountains or in taiga the average is $n \cdot 10^{-7}$ g/liter. Each figure represents the "background" content for that particular landscape and climate. Therefore, a sample running $n \cdot 10^{-5}$ g/liter may indicate a uranium mineralization in the mountains with a humid climate but not in a desert.

Generally, the aqueous dispersion haloes are smaller in deserts because metals tend to precipitate as their concentrations rise in the evaporating solutions. This detracts from the value of hydrogeochemical prospecting in deserts for such metals as copper.

THE SODA PROCESS

Soda waters. In the earth crust, the amount of strong cations (Ca, Mg, Na, K) is greater than the amount of strong anions (Cl, S, P, V). Therefore, more cations go into solution during weathering than do anions. The anion deficiency is compensated for by the dissociation of water. This dissociation proceeds in such a way that the number of hydroxyl ions increases as long as there is a demand for them and the number of hydrogen ions correspondingly decreases.

Since natural waters always contain some carbon dioxide, these waters gradually acquire sodium bicarbonate and an alkaline reaction. In other words, such waters gradually become "soda waters."

Soda waters are formed during the weathering of feldspars, which are most abundant in the granitoids (among igneous rocks) and in the gravels (among the sediments).

Egorov (1961) has drawn attention to the role of a sorbent in the removal of soda from water. According to him, migrating soda waters exchange some of their ions with calcium-bearing clays:

$$Clay = Ca + Na_2CO_3 \rightarrow Clay = 2Na + \underset{\downarrow}{\underline{CaCO_3.}}$$

The sodium ions are sorbed by the clay while calcium carbonate precipitates, making this clay calcareous. Thus, soda waters should be stable while in sands and sandstones unless these contain colloids in significant amounts. Conversely, soda waters are unstable in calcium-rich argillaceous sediments.

Egorov observed that the southern slope of the Tarim Basin (the part formed by Kunlun Mountain) is definitely contaminated with soda, while the northern slope (that of Tyan-Shan Mountain) shows a very small soda content in the soils. In his opinion, these differences are conditioned by climate and geology.

The cliff-studded Kunlun is in an extremely arid climate. Physical weathering predominates and the material, carried off by streams, is coarse and poor in colloids. All this makes exchange reactions difficult. As waters evaporate, soils become impregnated with soda.

The foothills of Tyan-Shan are composed of argillaceous rocks which contain gypsum. Consequently, the northern slopes of the Tarim Basin are not suitable for the accumulation of soda. In the soil itself, soda is generated during the exchange reactions. According to Gedroits (1933), the reactions of soda formation are:

$$\underset{\text{solid}}{\underline{soil = 2Na}} + \underset{\text{dil. sol.}}{\underline{Ca^{2+} + 2HCO_3^-}} \rightleftarrows \underset{\text{solid}}{\underline{soil = Ca^{2+}}} + \underset{\text{dil. sol.}}{\underline{2Na^+ + 2HCO_3^-}}$$

$$\underset{\text{solid}}{\underline{soil - Na}} + \underset{\text{dil. sol.}}{\underline{H^+ + HCO_3^-}} \rightleftarrows \underset{\text{solid}}{\underline{soil - H^+}} + \underset{\text{dil. sol.}}{\underline{Na^+ + HCO_3^-}}$$

As a result of such exchange reactions, soda is formed in aquifers, provided that these contain complexes capable of sorbing sodium (i.e., the "marine type" of complexes).

Soda can also form in the Hilgard reaction:

$$Na_2SO_4 + CaCO_3 \rightleftharpoons Na_2CO_3 + CaSO_4.$$

In other words, during the percolation of waters through limestones they precipitate gypsum. To prevent the reversal of the reaction, the inflow should be continuous.

In volcanic regions, sodium bicarbonate is formed in the alkaline hydrotherms which occur on the periphery of volcanic activity. These waters also carry silica and rare metals. Coming out on the surface, they form lakes and marshes. The well-known example is Searles Lake in California, which contains enormous amounts of trona, boron, lithium, and tungsten.

Within the supergene zone, waters are just as abundant on the surface as they are in the catagenetic subzone. Under conditions in humid climates, their mineralization is so low that neither the pH nor the other properties indicate the presence of sodium bicarbonate.

In deserts, the evaporation of waters is too rapid to stabilize soda waters. As the concentrations rise, the metamorphism of the waters is speeded up. Waters become sulfatic at first and then chloridic.

The most favorable conditions for the evaporation of ground soda waters are those of semiarid climates — steppes, forested steppes, subtropical and tropical savannahs. A number of artesian basins of the USSR are characterized by alkaline soda waters — in Tashkent, the Caucasus, and western Siberia.

Soda epigenesis. Soda waters condition some very peculiar rock alterations which may be termed "soda processes." Silica dissolves and migrates easily in such waters. Forming colloidal solutions of sodium humate, humic acids migrate easily. Aluminum forms the soluble sodium aluminate. Molybdenum also moves freely. A large group of elements which are difficultly soluble in neutral and weakly alkaline media become mobile due to the formation of water-soluble complexes.

The importance of this particular form of migration is stressed by Shcherbina (1956). For example, copper, which precipitates from neutral or weakly alkaline solutions in the form of malachinite and azurite, produces highly soluble complexes. The copper phosphate ellite is also soluble (Shcherbina and Ignatova, 1955).

Hexavalent uranium, which otherwise forms the insoluble carbonate ruthefordite produces in soda waters soluble complexes of the $[(UO_2)(CO_3)_3]^{4-}$ type (Shcherbina, 1957). It is a well-known fact that uranium cannot be precipitated in the presence of sodium carbonate.

Analogous complexes are also formed by silver, scandium, beryllium, zirconium, yttrium, and the yttrian rare earths. For example, zirconium forms the soluble complex $[(ZrO)(CO_3)_2]^{2-}$ (Degenhardt, 1959).

Therefore, a large group of elements can be extracted with soda waters. The following criteria are useful for the recognition of this phenomenon:

1. Silicated rocks and wood. The silication of a tree trunk is obviously accompanied by the copious evolution of carbon dioxide. Waters around a rotting tree stump become charged with carbon dioxide. Their pH is somewhat lowered (to 7.0-6.5?). Calcium carbonate begins to dissolve and silica to precipitate. The outcome is known as the petrification of wood. Various grains of quartz, feldspars, and other silicates are corroded at the same time.

2. Simultaneous migration of silica and alumina results in the formation of such aluminosilicates as palygorskite, attapulgite, and chrysocolla. Secondary albitization of rocks is another phenomenon.

3. Precipitation of brucite, $Mg(OH)_2$, from strongly alkaline solutions.

4. In silicated rocks, including those with silica concretions, a number of chemical elements, brought in soda waters, precipitates wherever humus is present.

The gypsification of rocks prevents the development of the soda process:

$$Na_2CO_3 + CaSO_4 = CaCO_3 + Na_2SO_4.$$

Gradually, soda waters are converted into sodium sulfate waters. However, if the soil or the rock contains just enough gypsum to offset this tendency, the sodic character of the water will be preserved. In such cases, gypsum is pseudomorphously replaced by calcite.

Many of the indicators of the soda process can be readily identified in thin sections. These indicators include the corrosion of quartz, silication, albitization, the existence of secondary clay minerals, and the pseudomorphism of calcite after gypsum.

The process of the formation of soda waters, which is going on in Tanganyika at the present time, and the resulting rock alteration have been described by Bassett (1954). That dry tropical climate of a forested savannah promotes the formation of soda waters during the weathering of igneous rocks. Silica, leached out during the weathering, is precipitated on the surface, causing the superficial silication of rocks. Soda lakes, like the Rudolf lake, are numerous.

It is very probable that soda waters predominated at one time in the Kalahari, Namib, and other deserts of southwest Africa. In western Bassarga (Iran), such phenomena have been recognized as long ago as the turn of this century. The formation of pelicanites (opalized kaolin rocks) and other types of silication may well be the result of the same soda process. Pelicanites are common in the Ukraine and Kazakhstan (Ginzburg and Rukavishnikova, 1952; Chukhrov, 1955). Presumably, pelicanites form at first as kaolin crusts and then become opalized during the change from a humid to an arid subtropical climate.

Some copper deposits may develop during soda catagenesis, e.g., the cupriferous sandstones of Dzhezkazgan, north-central Kazakhstan, northern Kirghizia, and the Donbass (Perel'man, 1959; Perel'man and Borisenko, 1962).

The rhodusite mineralization in marls of the Lower Permian period in the Dzhezkazgan Basin was probably the product of soda catagenesis.

Andreev and Godovikov (1959) hold that the most productive levels are gypsiferous lagoon beds.

Rhodusite, a fibrous alkali amphibole, is found lining the solution cavities in marls. Pseudomorphs of various carbonates and rhodusite after gypsum are not uncommon. Albite is the late mineral in these cavities. Rhodusite also occurs in the form of narrow veins, stringers, and disseminations. Its veins are usually opalized. Andreev and Godovikov believe the rhodusite mineralization to have been syntectonic.

The tectonic movements, which occurred after the deposition of the Permian strata, apparently rejuvenated the activity of the ground waters. Percolating through the sandy aquifers, the waters gradually acquired the sodic composition. The climate must have been arid.   Gypsum was pseudomorphously converted into calcite. At the same time, the waters became increasingly sulfatic. A decrease in alkalinity favors the precipitation of silica and alumina and, therefore, the synthesis of rhodusite. Sodium and magnesium must have been derived from the water. Bivalent iron was obtained from the host rocks.   All this probably happened at great depths and at rather elevated temperatures.

Chapter 11

The Gley Processes of Reduction in Epigenesis

The gley* forming processes are very common in soils, crusts, alluvium, and catagenetic subzones. Gley is formed in waters containing little, if any, hydrogen sulfide and no oxygen. Anaerobic bacteria thrive in such environments. They obtain their oxygen by breaking down different substances. Their waste products — carbon dioxide, methane, organic acids, etc. — are dissolved in water.

Iron and manganese are reduced to the bivalent state and become highly mobile — like any bivalent element. In acid media, trivalent iron and aluminum migrate only sluggishly if in combination with humic compounds. The redox potential may be negative or positive, rising as high as +0.4 V.

The mechanism of gley formation can best be observed in swamps which are being drained. The mud, scooped from the bottom of drainage ditches, is dark bluish to greenish. On drying, it oxidizes quickly and turns rusty red-brown due to the formation of trivalent iron oxides.

The bivalent, ferrous iron combines with the compounds of aluminum and silicon producing ferroaluminosilicates, these are the secondary clay minerals. They are greenish or bluish. The colors of the freshly exposed gley layers are usually due to such minerals.

The criteria for the recognition of gley processes are the hydroxides of iron and manganese which form as the redox potential rises and later become oxidized. Vivianite, $Fe_3(PO_4)_2 \cdot 8H_2O$, is also characteristic in environments free from carbonates. Humic substances and secondary clay minerals have already been men-

*This is a Ukrainian word for a greenish to bluish mud accumulating at the bottom of lakes and streams.

tioned. The process involves the colloidal migration and the colloidal sorption of copper, nickel, cobalt, silver, etc. As a rule, the gley strata become impoverished in iron, manganese, and phosphorus.

Gley waters. This type of water, containing no oxygen, occupies a special place in the geochemical classification of waters. More will be said about this in Chapter 15. Gley waters are characterized by a somewhat higher iron content and, locally, by higher contents of manganese, methane, humates, phosphorus, or copper. In turn, gley waters are typical swamp waters in the taiga and in the tropics. On the Siberian plain, the content of bivalent iron reaches 10 mg/liter. Since many artesian waters carry significant amounts of ferrous iron, they must belong to this type. For example, ground waters of the Poltava aquifer of the Dnepr artesian basin contain up to 28 mg/liter iron. The iron content of gley water in the west Siberian artesian basin varies between 20 and 30 mg/liter and occasionally reaches 100 mg/liter (Bogomyakov and Nudner, 1963).

In the Carpathian Mountains, gley waters commonly carry iron and the bicarbonate of calcium when in Cretaceous or Neogene rocks. In addition, such waters also carry Ba, Sr, Mn humates, much carbon dioxide, and, occasionally, hydrocarbons (K. P. Gayun). The bicarbonatic waters of Darasun in the Baikal region and in other folded regions belong to the same type, since they carry the alkaline earths and iron.

Depending on the intensity of mineralization, gley waters may be classified as ultra-fresh, fresh, saline, or briny.

Gley developing in the absence of carbonates is common in the tundra, the northern taiga, and, generally, in all permafrost regions.

The abundance of water in soils and rocks, combined with the humidity of the climate and slow evaporation, precludes the rapid and complete decomposition of vegetation remains.

Microorganisms progressively break up organic matter through intermediate products of the humus type into carbon dioxide and water. Since any oxygen which might have been dissolved in water is used up in the oxidation of organic matter, the waters have a reducing character. Soils and rocks assume bluish-gray colorations. The migration of manganese is marked. Waters range from weakly acid to neutral.

Gley forming in the presence of carbonates develops in soils and rocks containing calcium carbonate or dolomite.   Usually, waters range from alkaline to neutral, and more rarely are weakly acid. Calcium carbonate is both leached and redistributed by such waters. Manganese migrates relatively easily, precipitating in blocks with higher oxidation potential. Dendrites and stringers are formed by black manganese hydroxides.

This type of gley is common in swamps developed over carbonate rocks. The most favorable climates are the humid ones, from that of the tundra to that of the forested steppes. Occasionally, gley develops in semideserts and even in deserts.

Forming in the presence of gypsum or other salts, gley is conditioned by waters carrying those salts in solution. It may develop in the subsurface as well as in swamps (Perel'man, 1959, 1961).

Gley formation in red rocks takes place in rocks of very different composition: gravels and sandstones, alevrolites (cemented clastics), and clays—provided that such strata alternate. The predominant color of the rocks is red, due to the films of ferric hydroxides which develop over the individual grains of sand, silt, or clay.

Bleached spots and bands of bluish or greenish white stand out in contrast with the general red color of the oxidized sandstone, conglomerates, and dolostones. Such series of varicolored layers are typical of the Paleogene, Cretaceous, and Lower Carboniferous rocks in Central Asia and of the Permian rocks in the Donbass and the Ural region.   In Kazakhstan, the red rocks are Carboniferous. In other words, the phenomenon has a regional significance.

We interpret these varicolored bands as ancient water courses which carried deoxygenated waters at one time or another. It is only natural that such discolorations are restricted only to the permeable strata.   Carrying much carbon dioxide and being reducing toward iron, those waters removed iron as iron bicarbonate and formed the gley. As a result, the rocks acquired greenish- to bluish-gray colors, except where the formation of the ferroaluminosilicates prevented the removal of iron. Due to capillary sorption, the adjacent impermeable strata became wet along the contacts with the aquifers, lost some of their iron, and became impregnated with gley. These portions also became discolored.

Let us consider the following cases in order to appreciate fully the morphology of the gley emplacement. The first is the case of gley formation in the absence of carbonates cited in the Zerabulakh Mountains in the Samarkand State of the Uzbek SSR. The Senoma-

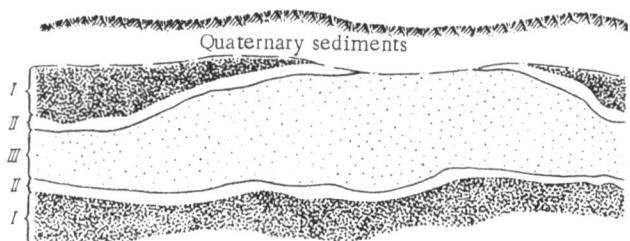

Fig. 41. Epigenetic profile of Senomanian red rocks (at the base of U. Cretaceous). (I) Impermeable clay, syngenetically colored raspberry red; (II) impermeable colloidal clay, syngenetically colored greenish; (III) water-bearing sandstone.

nian red rocks exposed on the steep side of a dry gulch are (Fig. 41):

Quaternary.	Gravels with brown sand and clay, with pebbles of the Paleozoic rocks	0.00 m
Senomanian.	Raspberry-colored alevrolites	2.50 m
	Bluish-gray to green alevrolites, irregularly passing into the raspberry-colored ones.	3.00 m
	Gray sandstone with small pebbles of the Paleozoic rocks; locally brown or bleached white.	3.20 m
	Bluish-gray to green alevrolites, identical with those above and grading irregularly into the raspberry-colored ones.	4.10 m
	Raspberry-colored alevrolites	4.20 m
	The end of this series.	5.20 m

We interpret this geological section as consisting of an old water course, the sandstone bed enclosed between water-tight alevrolites. The ground waters, which moved along that water course at one time, carried no oxygen. The leaching of iron and the emplacement of gley extended into the adjacent alevrolites irregularly along their contacts.

The formation of gley in the presence of carbonates is represented by a Lower Cretaceous section in southern Fergana (Fig. 42). That section was measured on the southern slope of the Guzan anticline just north of Isfarah. It consists of intercalated clays, alevrolites, sandstones, and conglomerates. Generally, the color of the series is red with spots and laminae of bluish-gray to green color. The development of the epigenetic gley was limited mostly to sandstone and conglomerate members. Beds of less than 1 m in thickness were completely impregnated with gley. Thicker strata carry gley only near the contacts with clays and alevrolites. These clays and alevrolites are irregularly impregnated with gley to a fraction of one meter from the contact. The middle part of the Muyansk series is:

Red-brown clay, locally raspberry colored, crumbling into angular fragments which are coated with gley. . . . . 0.00 m

Bluish-gray to green clay, the product of alteration of the raspberry-colored clays; the same texture but with inclusions of red relicts; somewhat spotty but generally bluish-gray to green color. . . . . . . 5.00 m

Gley-impregnated sandstone, bluish gray with red-brown, rusty spots. . . . . 5.20 m

Red-brown alevrolite with gley lining the fissures. 6.20 m

The end of this series. . . . . 7.00 m

All members of this section effervesced with the addition of hydrochloric acid.

It was concluded that the waters which percolated along that water course carried no oxygen. Since no gypsum was found, the waters were very slightly mineralized; since calcium carbo-

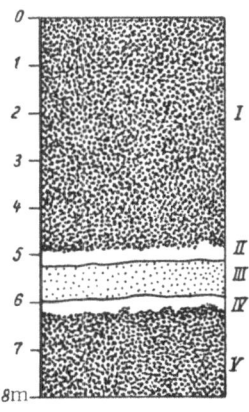

Fig. 42. Epigenetic profile of red rocks of the Muyansk series (at the base of U. Cretaceous of S. Fergana). (I) Raspberry-colored clay; (II) colloidal clay; (III) bluish-gray sandstone, spotted with ocher; (IV) colloidal alevrolite; (V) red-brown alevrolite.

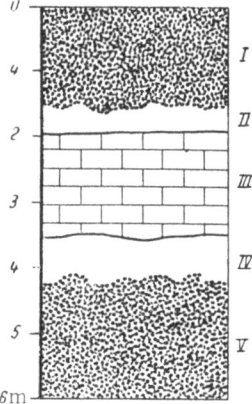

Fig. 43. Epigenetic profile of red rocks of U. Cretaceous on the Sokh River. (I) Red alevrolite; (II) green alevrolite; (III) limestone; (IV) colloidal alevrolite; (V) red alevrolite.

nate was present, the waters must have had a neutral to weakly alkaline reaction.

Another interesting case was seen by us in the Shorbulak region, some 10 km east of the city of Erevan. The sandstones and conglomerates were impregnated with gley and thus had a green color. The intercalated red clays contained no gley and vigorously effervesced with the addition of hydrochloric acid. The sandstones and conglomerates generally did not react with the acid except in narrow bands (10 to 20 cm) along the contacts with clays. These border bands effervesced moderately with the acid, and, in addition, were spotted with iron ochers.

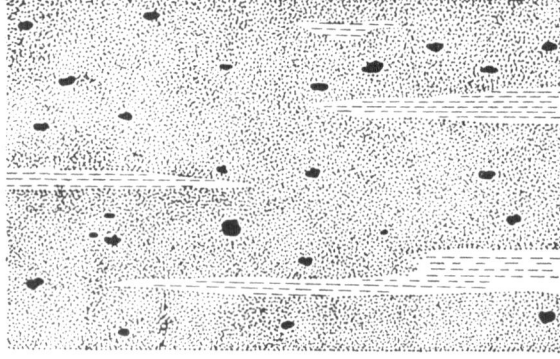

Fig. 44. Emplacement of colloids in lenses of clays and alevrolites included in gravels.

The same phenomena can also be observed near the village of Chastyie in the Kama River valley. The sandstones of the Ufa Red Rock series may have carried very little calcium carbonate or perhaps none at all. This indicates that the waters carried bicarbonates and probably had a rather low pH (6-6.5?). The pH probably was higher near the contact with calcareous clays, and some calcium carbonate precipitated in the border bands. On the whole, however, the conditions were not favorable for the formation of gley within clays, and these retained their red color. Obviously the pH of the media fluctuated about the "border-line," since even the slightest rise in pH would produce the gley. The precipitation of ferrous iron is possible near the contacts of carbonate and carbonate-free strata. It precipitates in the form of siderite, and, eventually oxidizing, produces some reddening of the rock. Thus the gley can conceivably form on both sides of the boundary between the oxidizing and the reducing media.

In the catagenetic subzone, the gley may form in any aquifer, including the fissured limestones. This is shown by the geologic section measured in the Cretaceous red rocks near Kalach on the Sokh River (Fig. 43). The iron removed from the alevrolites during their impregnation with gley was reprecipitated in limestones in the form of ocherous impregnations near their contacts. A similar situation was noted near Sarakamysh on the Sokh River, where alevrolites underlie the limestones.

As a rule, the red rocks do not contain sufficient organic matter to reduce all the ferric iron present in an aquifer. The reduction of ferric iron is spotty, accounting for the mottled colors of bluish gray to green. The occluded red ferric hydroxides are relicts. Red-brown ferric hydroxides were deposited by ground waters. In spots, however, the gley reaction has been completed.* It is interesting to note that the formation of gley may be on a large or a small scale. For example, thin stringers (2 cm) of bleached material frequently occur in brown conglomerates and sandstones (Fig. 44).

In all probability, the waters circulating in these strata were oxygenated. Such strata retained their original red or brown color while the color of the enclosed clays and alevrolites changed

---

*It does not follow from the foregoing that all the gray layers in the red rocks have been produced epigenetically. Some may have undergone syngenetic reduction, i.e., during their deposition. Rocks rich in organic matter, such as marls, clays, and alevrolites, are possible examples.

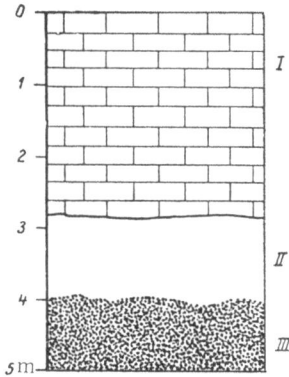

Fig. 45. Emplacement of colloids in sarmat clays (base of U. Miocene), at Ust'-Urt, Uzbekistan. (I) Sarmat limestone; (II) colloidal clay; (III) red clay.

due to the stagnation of the waters there. The leached iron, reduced to the bivalent state, diffused in the direction of the circulating waters. Then, mixing with oxygenated waters, the iron oxidized and precipitated. Such was the probable origin of ferruginous sandstones occurring in the varicolored formations. It has been established that the bluish-gray clays are poorer in iron.

Another possibility is that the ferruginous impregnation began in a reducing environment as the deposition of siderite. Later, the oxidation of siderite resulted in the accumulation of ferric hydroxides (see below).

The geochemical activity of ground waters may have been so vigorous in places and the number of water courses so great that the iron films were dissolved and removed. In any case, the strata acquire a mottled, spotty appearance. The bleached members appear sandwiched in between heavily ferruginous beds.

Iron oxides and hydroxides occurring in the red rocks are of various shades of red, brown, or yellow. Red colors are specific of hematite and of those oxides which contain relatively little water. Such colors are syngenetic, for the red oxides and hydroxides were formed on dry land under conditions of subaerial weathering in hot, arid climates. On being eroded, this material may redeposit in river deposits.

In the catagenetic subzone, the red rocks are partially reworked by ground waters. This results in a redistribution of iron and the formation of gley. Ferric hydroxides contain much water when precipitated from ground waters and display yellow to brown colors. However, the colors may be also imparted at a later date; the reddening of outcrops is an example.

Although most common in aquifers, gley may form under different conditions.

The formation of gley in other rocks. The catagenetic formation of gley takes place in various rocks, although the phenomenon is never as clearcut as in the red rocks.

The carbonate-free gley formation is well marked in coal veins of Jurassic age in Central Asia. These suites consist of alternating alevrolites and ferruginous sandstones. Similar phenomena have been observed in the Lower Cretaceous sands north of Moscow and in the sands and opalites of the Volga uplift. In all cases, the gley impregnation is confined to the ancient water courses.

Gley formation is also common in calcareous rocks. We observed an interesting case at the southern tip of the Kaiłak marsh on the Ust'-Urt plateau. There, the Sarmat limestones are intercalated with bright red clays which are impregnated with gley along contacts (Fig. 45).

This particular section may be interpreted on the basis of geography and the present and past fluctuations of the ground water table. Trapped in the Sarmat limestones, the ground waters were sorbed by the intercalating clays and eventually became cemented with gley. The evaporation of water was probably vigorous in the aquifer's upper part and the overlying soils became contaminated with the evaporated salts. Thus, gypsum was being added to the soil as the gley impregnated the clays underneath.

It follows that gley layers can be used in the reconstruction of paleogeographic conditions.

In the Ukrainian lowlands, which border on the Black Sea, Pontian (L. Pliocene) limestones are not infrequently overlain by a so-called Scythian clay of bright red color. The gley impregnating these clays near their contacts is paragenetic.

Near the village of Krinichki on the Bessarabian coast, Pontian limestones are spotted with rusty brown (Fe) and black (Mn) where exposed in a quarry. They are overlain by three to four meters of Scythian clays which are green in the lower 1.5 m and gradually change to red-brown in the upper part. The lower part was obviously cemented with gley deposited by waters rising from the limestones underneath. Since the clay was laid in an arid climate and became gley impregnated only during catagenesis, the red-brown color of the upper part is not due to the weathering of the

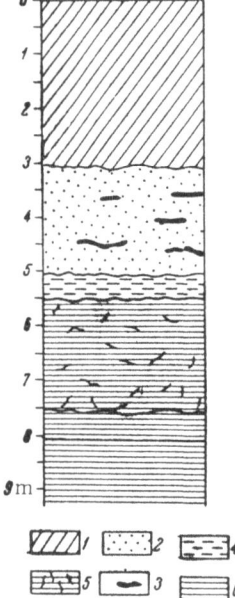

Fig. 46. A geologic section near the village Yaloveny. (1) Loess; (2) E. Quaternary sands; (3) iron impregnation; (4) colloidal clays; (5) brown-green oxidized clays; (6) gray sarmat clays.

green material.   The brown and black spots along the fissures were formed during the gley emplacement.

Gley formation in the carbonaceous shales. Black carbonaceous shales are widely spread throughout the Soviet Union. They are mostly Paleozoic. Carbonaceous and siliceous shales of the Tyan–Shan region are examples. On their surface, such shales may be intensely altered and varicolored.   The bleached and the impregnated blocks alternate with the ferruginous or black (unaltered) blocks.   The familiar lithologic control of color is seen everywhere.   The sandy lenses are impregnated with iron ochers, while the clayey lenses are bleached. Occasionally, some black shale has been preserved between two ancient aquifers.

Since this alteration extends laterally and not downwards, it cannot be ascribed to the weathering.   Even on the surface outcrops, bleached strata may lie sandwiched between unaltered ones. Furthermore, the weathering in deserts, where these phenomena are encountered, produces carbonates and not these alterations.

The bleaching of some carbonaceous shales may well be due to the activity of microorganisms feeding on the organic matter contained in shales.

Altered shales stretch in wide belts for many kilometers. For example, the belt of varicolored shales of the Sughet series (Silurian) can be seen on exposures along the Shakhimardan and Sokh Rivers in the southern Fergana. The same bleached and impregnated shales can also be seen at the Beshbulak well at Kyzylkum, about 45 km northwest of Tamda.

It is interesting that in the southern Fergana the gley-bearing Sughet shales are directly overlain by the gley-bearing red rocks of the Lower Cretaceous. We believe that the gley impregnation of both rocks could not be older than the Lower Cretaceous.

Gley impregnation of red rocks and oxidation of gray rocks. The oxidation of gray pyritic clays results in the formation of varicolored suites which display rusty brown spots of limonite on the general green of the background. Outwardly, such oxidized clays resemble the gley-impregnated ones. Occasionally, these rocks may become gley impregnated.

An exposure in a limestone quarry near the village of Yaloveny south of the city of Kishinev is reproduced in Fig. 46. The topmost loess is underlain by cross-bedded sands of the Lower Quaternary period which occlude lenses of a coarser material. These sands in turn are underlain by green clays of the Middle Sarmat age, which are spotted with ocherous brown. At greater depth, the color of these clays changes to a solid gray. Stringers of brown ferruginous and black manganiferous material occur in the sands. The chronological order of events was as follows: (1) deposition of gray Sarmat clays in reducing environments; (2) weathering and oxidation of this clay; (3) burial of these clays under the Quaternary sands and loess; and (4) the development of water courses; gley and leaching of Fe and Mn. Consequently, the green clays with brown spots have been undergoing an epigenetic alteration. Their uppermost portion was receiving gley, their middle portion was oxidizing, and their lowermost portion was reducing. Morphologically, the middle portion resembles the uppermost very closely.

Epigenetic sideritization. The formation of epigenetic siderite is closely related to the formation of gley. Siderites were formed in the ancient water courses whenever the carbonate equilibrium became disrupted:

$$Fe\,(HCO_3)_2 \rightleftarrows FeCO_3 + H_2O + CO_2.$$

This happened at Kyzylkum, where Eocene conglomerates often contain a limonitic cement which occludes siderite relicts.* The underlying red Eocene clays are impregnated with gley on their contacts with the conglomerates, their color in such places being green. Iron, removed from clays during that impregnation, was transferred to conglomerates and redeposited there as siderite. The subsequent uplift and exposure of the conglomerates to weathering resulted in the oxidation of siderite and the formation of brown ironstones. Similar phenomena have been observed in the Senoman conglomerates in the Zerabulak Mountains in Central Asia.

According to Kuzemkina (1962), the emplacement of gley and siderite is typical in many bauxite ore deposits. In Mesozoic rocks of the western Turgai, the epigenetic bleaching of bauxites is conditioned by ground waters and lignitic clays. As the gley develops in one, iron is leached out of the other and is eventually converted to siderite.

Similar processes of the gley-siderite formation have produced certain red soils over the Pontian limestones in Moldavia. Near the village of Musaid some 18 km northwest of Bolgrad a bed about 2 m thick exists which is heavily ferruginous and bright red. This bed contains numerous fragments of white limestone embedded in a sandy clay. The fragments themselves are "dusted" on the periphery with the same red clays which impart a rose tinge to the white of the limestone. However, this sandy clay is neither the red soil nor the weathered crust. The bed still retains its stratification, but does not show the vertical breakup which is normally caused by vegetation roots. At the contact with this bed, the underlying sands still have a red color, but 0.5 m lower their red color changes to bluish gray, i.e., to the color of gley.

Therefore, we conclude that in the bed the Pontian rocks were probably saturated with gley-forming ground waters ever since their burial. These rocks might have become part of an artesian basin. The waters entrapped in the limestones were more dynamic than those of the sandstones. The pH was higher in the limestones than in the sands. This permitted the gley formation in sands, while the iron diffused into the limestones. It precipitated there as siderite because of the increase in pH and decrease in the

_____

*A good exposure of these conglomerates is found 3 km north of the Beshbulak well.

partial pressure of carbon dioxide. Later, this siderite layer was uplifted, oxidized, and converted into the red "pseudo-soil."

Emplacement of gley and kaolin. The oxidation of organic matter during the gley formation in carbonate-free rocks results in a buildup of CO_2 and humates in the water. The pH of the water drops and the weathering of feldspar intensifies and produces kaolin. White rocks develop which may consist of kaolin alone or kaolin mixed with quartz. Such epigenetic kaolinitization, contemporaneous with gley formation, is not a result of weathering, which would have oxidized the available iron and thus produced red or brown colors. The simultaneous formation of gley and kaolin is characteristic of regions which are lacking carbonates and rich in organic matter such as swamp bottoms.

Gley formation in fault zones. Gley formation is typical of fault zones carrying hypogene waters. Although derived from crystalline or other rocks deficient in organic matter, such waters are usually enriched in iron and carbon dioxide. This content suggests the gley origin of such waters.

Rusty-brown, ferruginous impregnations form in rocks around the places of emergence of such springs. In the Caucasus, the cold springs of this "narzan" type occur in the valley of the Baksan River and near the Crestovy Pass (on the Georgian Military Road). In Armenia, the Dilizhan springs should be mentioned. The travertine deposited by such springs is frequently red to orange in color due to iron impurities. However, should the waters be weakly acid, no carbonates would be deposited.

We thus, meet with the manifestations of gley formation whether the carbonates are present or not. Processes of this nature must have been very common in the past geologic epochs. In Central Asia, the black carbonaceous shales of the Cambrian, Ordovician, and Silurian periods have been impregnated with gley in the past, since they now lie above the ground water table.

The former layer of active water circulation, such as the fissured zones in slates, are frequently stained red or brown. The former zones of sorption — the adjacent shales — are whitened by the deposited kaolin. The surface outcrops of such rocks may be mistaken for varicolored suites because of the alteration of red and white beds.

Iron and manganese impregnations in fissures. Some staining of fissure walls with iron or manganese is common in all mountainous regions. At the same time, the nonfissured rocks retain their gray colors.* This ferruginous impregnation cannot possibly be ascribed to the effect of modern weathering, since the oxidation of pyrite would have affected both the shales and the sandstones. In fact, the shales should have been oxidized to a greater extent because of the greater pyritic content in the marine shales. Nevertheless, the sandstones became colored yellowish-brown while clays and shales remained gray. The color change is not gradational but sharp, as if a ferruginous film were super-imposed onto the fresh surface of the rock.

We presume that the ferruginous and manganiferous films and smudges are, in many cases, the leftover traces of the gley reaction which developed when the fissured rocks were still below the erosional basis. After the uplift the gley waters began to oxidize. There are several possibilities of what happened: Iron could have precipitated while manganese remained in solution, both elements could have precipitated at the same time, or manganese alone could have precipitated (the iron having fallen out at lower depths). The films could have been produced by different factors. In some cases, there should have been a definite sequence whereby the zone of manganese films was underlain by a zone of ferru-ginous films.

The following case confirms our conclusions. In Armenia, basalt flows are numerous but none is favorable for the formation of gley ground waters. An exposure of Quaternary basalt in a quarry some 7 km northwest of the village of Sisian on the road to Bazarchai-Goris is typical. Under a fairly shallow crust, basalt is fractured. These fractures show no alteration of the walls. Had the weathering been responsible for ferruginous impregnations in fissured rocks elsewhere, the same could be expected here. Ferruginous impregnations have been found in the older basalts and porphyrites of the same region, near the Chaikend bridge on the Arpa River.

Ferruginous and manganiferous impregnations and incrusta-tions in fissures are remnants of ancient hydrochemical zones which developed below the regions of oxygenated waters. Old hydrochemical zoning can be reconstructed on the basis of rock alteration.

---

*On the southern slopes of the Greater Caucasus, there are outcrops of fissured yellowish-brown sandstones intercalated with gray argillaceous marls. When freshly broken, these sandstones display gray colors in their interior. Similar ferruginous oxidation and impregnation along fissures have been observed in many other regions of the USSR.

It does not follow from the foregoing that every ferruginous deposit in fissures was caused by the oxidation of gley waters. Fissures may be incrustated or even filled in during the course of pyrite oxidation.

Hydrothermal gley. Springs issuing waters with a high content of Fe^{2+} are well known among the hot mineral springs dotting the alpine-type mountainous regions. The ground around their points of issue is stained various shades of yellow, reddish-brown, and orange by the precipitating iron compounds. As an example, we might mention the thermal (54° to 64° C) springs of the Dzhermuk resort in Armenia. These springs deposit reddish and yellow travertines.

Such phenomena point to a hydrothermal impregnation with gley at depths below those of cold water. Many cases of bleached rocks in fault zones should be due to hydrothermal gley formation.* Contemporary thermal waters are known to exist in geosynclinal regions as well as in stabilized platforms. The already cited springs of Dzhermuk are examples of the first type. The second type is represented in the Siberian lowlands, where gley waters have temperatures of 50° to 70° C at depths of 1000 to 1500 m (Zaitsev and Tolstikhin, 1960).

Migration of chemical elements during gley formation. The chemical analyses of the raspberry-colored and the bluish-gray alevrolites are given in Table 29. Obviously, iron, manganese, and phosphorus have been removed during gley formation in the carbonate-free environment. Comparing the analyses of the unaltered rocks and their gley-impregnated derivatives, it is difficult to say what was removed and what was added. Any conclusions reached would have to be purely qualitative.

A method of quantitative evaluation of what was removed and at what rate was first proposed by Merrill. Let us consider its application to the Senomanian alevrolite. From the data of Table 29 we select an "oxide-indicator" which neither increased nor decreased in amount during the alteration. Alumina, Al_2O_3, was chosen as an indicator in the evaluation of the samples 412 and 411. It will be noted, however, that the analysis of the gley product yielded 13.37% Al_2O_3 instead of the theoretical 12.01%. The difference is the apparent enrichment due to the residual concentration of alumina left after some other components were removed. It can be calculated from these data that only 89.84 g are

*More about it in Chapter 15.

Table 29. The Gley Impregnation of Albian-Senomanian Alevrolite in Carbonate-Free Environments (Zerabulak Mountains, Uzbekistan)

Sample no.	Rocks and their factors	Content, %								
		SiO_2	TiO_2	Al_2O_3	Fe_2O_3	FeO	MnO	MgO	CaO	Na_2O
412	Raspberry-colored alevrolite, a	73.67	0.53	12.01	3.57	None	0.02	0.70	0.25	0.85
411	Green alevrolite, gley impregnated, b	73.04	0.80	13.37	1.72	None	Tr	0.86	0.01	0.78
	b	65.74	0.72	12.01	1.54	-	-	0.770	0.009	0.76
	b − a	-7.93	+0.19	0.00	-2.03	-	-	+0.07	-0.241	-0.02
	Relative change $(b-a)/a \cdot 100\%$	-10.76	+3.58	0.00	-56.86	-	-	+10.0	-96.4	-2.56

Sample no.	Rocks and their factors	Content, %								
		K_2O	P_2O_5	Cl	SO_3	H_2O^-	H_2O^+	CO_2	Loss on ignition	Total
412	Raspberry-colored alevrolite, a	1.47	0.03	0.35	0.19	1.07	4.39	0.48	0.30	99.81
411	Green alevrolite, gley impregnated, b	1.40	0.02	0.25	0.52	1.64	4.32	0.66	0.49	100.04
	b	1.26	0.018	0.22	0.56	-	-	-	-	-
	b − a	-0.21	+0.012	-0.13	+0.37	-	-	-	-	-
	Relative change $(b-a)/a \cdot 100\%$	-14.28	-40.0	-37.14	+194.73	-	-	-	-	-

left now from every 100 g of the rock as it was before the alteration:

$$\frac{100 \text{ g} \cdot 12.01 \text{ g}}{13.37 \text{ g}} = 89.84 \text{ g}.$$

Multiplying the percentages of the oxides in analysis 411 by this
factor (0.8984), we obtain the data on line "b." The difference, $b-a$,
is the increment (positive or negative) of the given oxide during
the gley reaction. The ratio $(b-a)/a$ is the measure of gain or
loss.

The gley emplacement was obviously accompanied by a vigorous
removal of calcium, iron, and phosphorus (Table 29). Silica,
sodium, and potassium were removed to a lesser extent. Sulfur
was probably added afterwards. In this case the migration series
is:

$$Ca > Fe > P > Cl > K > Si > Na.$$

Different elements become mobile and again precipitate, depending on the changing redox potential. Thus, the reduction of
tetravalent manganese begins at relatively high Eh values when
the waters are still oxygenated and the iron is still trivalent.
The reduction of ferric iron starts as the Eh lowers. The reduced,
bivalent iron partly migrates and partly becomes fixed in the
form of leptochlorites or siderite. At this stage, copper and
uranium still remain oxidized at their higher valences, Cu^{2+} and
U^{6+}. As the Eh keeps dropping, the environment becomes favorable
for the precipitation of uranium as U^{4+} and of vanadium as V^{3+}.
Eventually, when the Eh has dropped enough, copper precipitates
as insoluble Cu^+ and $Cu°$.

The carbonate-free environments favorable to the formation
of gley can be divided into three groups:

1. Weakly reducing (Mn^{4+} to Mn^{2+}, Fe^{3+} to Fe^{2+}). These elements are partly fixed as leptochlorites and siderite.
2. Moderately reducing (U^{6+} to U^{4+}, V^{5+} to V^{3+}). Uranium and
   vanadium remain insoluble; iron and manganese migrate.
3. Strongly reducing (Cu^{2+} to Cu^+ and $Cu°$). Manganese and
   iron migrate, although iron may form secondary minerals.
   Uranium, vanadium, and copper remain insoluble.

It is of considerably practical as well as academic importance
to study what it is that produces the reducing environments. In

Table 30. The Gley Impregnation of Cretaceous Red Rocks in Carbonate-Rich Environments (Central Asia)

Sample no.	Rocks and their factors	Content, %						Soluble in 2% HCl					
		P_2O_5	SO_3	Cl	CO_2	Loss on ignition	Total	CaO	MgO	$CaCO_3$	$MgCO_3$	H_2O^-	H_2O^+
	Kughitang range, dolostone alevrolites of the Al Murad series (Valanzhin-Goteriv)												
122	Red alevrolite, *a*	0.11	0.20	0.02	19.08	22.88	99.80	12.90	11.24	23.09	16.98	—	—
120	Green alevrolite	0.14	0.85	0.47	12.16	16.92	98.47	8.14	17.80	14.57	10.94	—	—
121	Blue-gray alevrolite	0.14	0.59	0.75	12.20	17.06	99.62	7.89	8.70	14.12	11.40	—	—
	b	0.10	0.66	0.36	—	—	—	—	—	—	—	—	—
	b − a	−0.01	+0.46	+0.34									
	(*b − a*)/*a* · 100%	−9.9	+2.30	+1700									
	The intensity of removal: Ca > Mn > Mg > Fe > P > K > SiO_2												
	Red rocks of U. Cretaceous, River Sokh (Goznau)												
402	Red alevrolite, *a*	0.15	0,09	0.53	9,38	0.12	100.38	—	—	—	—	3.15	3.12
403	Green alevrolite, gley impregnated	—	0.24	0.33	5.40	0.07	100.43	—	—	—	—	3.17	2.17
	b	0.00	0.23	0.32	—	—	—	—	—	—	—	—	—
	b − a	−0.15	+0.14	−0.21									
	(*b − a*)/*a* · 100%	−100	+155,5	−39.5									
	The intensity of removal: P > Mn > Mg > Cl > Ca > Fe												
	Cretaceous red rocks, River Sokh (Vakhsh canyon)												
463	Raspberry-colored alevrolite	0.17	None	0.01	6.85	None	99.49	—	—	—	—	0.09	3.27
464	Gleyed alevrolite	0.12	0.13	0.04	11.72	—	102.22	—	—	—	—	0.65	2.58
	b	0.13	—	0.04	—	—	—	—	—	—	—	—	—
	b − a	−0.04	—	−0.03									
	(*b − a*)/*a* · 100%	−23.5	—	+300									
	The intensity of removal: Fe > Na > Ti > Si > P > K												

Table 30 (continued)

Sample no.	Rocks and their factors	Content, %										
		SiO$_2$	Fe$_2$O$_3$	FeO	Total Fe oxides*	TiO$_2$	MnO	Al$_2$O$_3$	CaO	MgO	K$_2$O	Na$_2$O
	Kughitang range, dolostone alevrolites of the Al Murad series (Valanzhin-Goteriv)											
122	Red alevrolite, a	34.44	2.51	1.30	3.94	0.45	0.11	9.44	12.90	12.85	2.22	0.57
120	Green alevrolite	43.50	3.10	0.50	3.65	0.60	0.08	12.07	8.14	9.95	2.65	1.35
121	Blue-gray alevrolite	44.26	3.39	0.35	3.78	0.60	0.06	11.26	7.89	9.80	2.65	1.41
	b	33.93	—	—	2.85	0.46	0.06	9.44	6.35	7.76	2.06	1.05
	b − a	−0.51	—	—	−1.09	+0.01	−0.05	0.000	−6.55	−5.09	−0.16	+0.48
	Relative change (b − a)/a · 100%	−1.16	—	—	−27.6	+2.2	−45.4	0.000	−50.7	−39.6	−7.20	+84.1
	Red rocks of U. Cretaceous, River Sokh (Goznau)											
402	Red alevrolite, a	49.75	4.76	None	4.76	0.56	0.14	11.35	—	5.70	3.89	
403	Green alevrolite, gley impregnated	59.54	2.80	0.48	3.33	0.74	0.06	11.72	—	2.96	4.45	
	b	57.75	—	—	3.23	0.71	0.058	11.72	—	2.87	4.32	
	b − a	+8.00	—	—	−1.53	+0.15	−0.032	0.000	—	−2.83	+0.42	
	Relative change (b − a)/a · 100%	+16.0	—	—	−32.1	+20.2	−58.5	0.000	—	−49.6	+9.43	
	Cretaceous red rocks, River Sokh (Vakhsh canyon)											
463	Raspberry-colored alevrolite, a	55.22	3.34	1.60	5.10	0.72	0.12	13.15	4.99	4.88	3.24	1.83
464	Gleyed alevrolite	48.34	0.82	1.86	2.87	0.59	0.16	11.82	8.79	8.70	2.51	1.35
	b	53.65	—	—	3.19	0.65	0.18	13.15	9.76	9.66	2.79	1.49
	b − a	−1.57	—	—	−1.91	−0.07	+0.06	0.000	+4.77	+4.78	−0.45	−0.34
	Relative change (b − a)/a · 100%	−2.84	—	—	−37.44	−9.72	+50	0.000	+95.7	+98.9	−1.38	−13.1

*The total Fe oxides are expressed as ferric oxide.

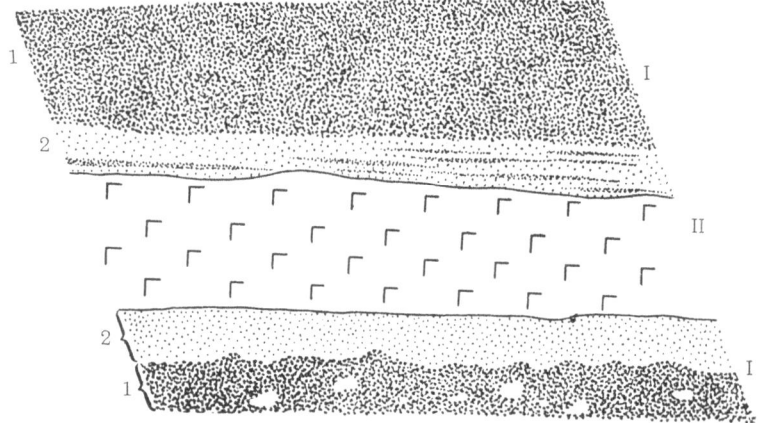

Fig. 47. Emplacement of colloids in red rocks containing intercalations. (1) Alevrolite:
(1) raspberry red syngenetic coloration; (2) greenish coloration of colloids. (II) Gypsum.

many cases, the reducing environments are conditioned by the
organic matter which became trapped during the initial sedimentation.   However, not all the gley-impregnated rocks contained
organic matter in the form of vegetal remains.   Gaseous hydrocarbons escaping from the oil and gas pools account for some
cases.   It is a well-known fact that oil and gas deposits occur
under the red rocks in Central Asia, the Ural Region, and elsewhere.   It has been proved that in geosynclines, the hydrocarbon
gases rise for hundreds of meters above the gas and oil pools.   It
is also well known that microorganisms feed on hydrocarbons
dissolved in waters.   Large reservoirs of natural gas have been
recently discovered in Permian red rocks on the Shebelinka,
Chervonodonets, and Spevakov structures in the Donbass (Kozlov
and Gladyshev, 1960).   These occurrences suggest that gley
impregnations may indicate a presence of gas or oil at depth.

Gley combined with oxysalts.   Desulfurization of
processes commonly develop in sulfate-rich rocks permeated with
chlor-sulfatic waters.

If hydrogen sulfide begins to generate, it limits the migration
of iron, copper, and other sulfides by precipitating these in the
form of insoluble sulfides.   The formation of gley is thereby prevented.   Nevertheless, under certain conditions the gley can form
in the sulfate-rich rocks.   We observed such gley impregnations
in the Upper Cretaceous red rocks in the Zeravshan River valley,
where the red alevrolites contained gypsum (Fig. 47).   We suppose

that the ground waters contained gypsum (its solubility is 2 g/liter). The Eh must have been too low to reduce the ferric iron and too high to start the reduction of the sulfate ion. The gley impregnation develops in sulfatic environments only if the desulfurization is precluded by the lack of organic matter or some other strong reducer.

The role of gley in ore deposition. A vigorous removal of iron and manganese during the development of gley in carbonate-free environments is coupled with the transfer of these elements to wherever the existing reducing conditions are changed again to the oxidizing. Thus, the blocks of gley-impregnated carbonate-free rocks are the probable sources of iron and manganese ores.

In swamps of Belorussia, the same process is apparently responsible for the formation of vivianites, which are used locally for fertilizer.

The role of the gley forming in carbonate-free environments may be instrumental in the formation of various ore deposits and thus needs further study.

The role of gley in carbonate environments was not determined. Our investigations indicate that cupriferous sandstones of the Ural Region, Central Asia, Donbass, and Dzhezkazgan are gley-impregnated layers of the red rocks. Much remains to be done to ascertain the role of gley in copper mineralization. Such gley strata may actually represent the buried dispersion haloes with the ore elements masked in hydroxides of iron and manganese.

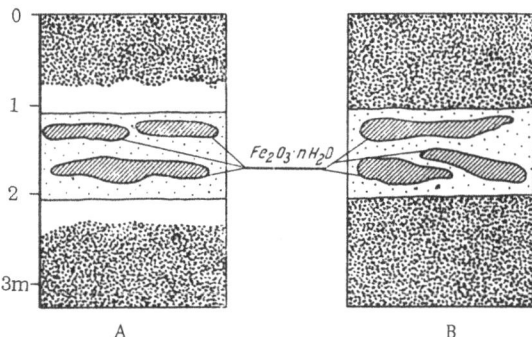

Fig. 48. Ancient colloidal emplacement due to demineralization of ground waters. (A) Oxidation; (B) iron impregnation.

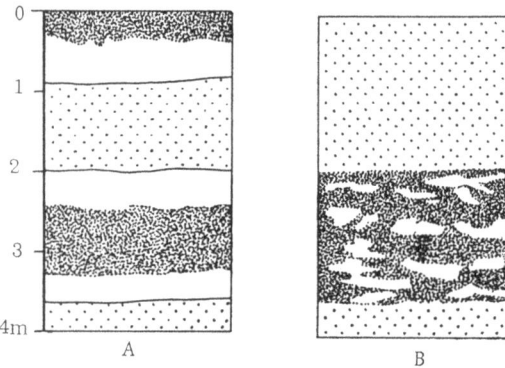

Fig. 49. Emplacement of colloids: (A) Linear catagenetic; (B) spotty diagenetic.

Analyses of such hydroxides may lead to a discovery of buried ore deposits.

The criteria for the recognition of gley, which may also be used for the reconstruction of paleohydrogeologic conditions, are: (1) the shoe-string strata in which bluish-gray layers are intercalated in ferruginous beds; and (2) films, smudges, dendrites, nodules, or concretions of hydroxides of iron and manganese, which are associated with accumulations of humus or other vegetal detritus.

The ferruginous impregnation of outcrops may be due to various causes. If ferruginous impregnations of water-permeable strata are caused by the present-day "unloading" of waters containing Fe^{2+}, there will be no gley-impregnated beds associated with the iron-impregnated ones (Fig. 48).

Diagenetic significance of the gley. Gley forms diagenetically as well as epigenetically. The diagenetic origin of gley is recognized by the occurrence of its spots scattered irregularly throughout the entire thickness of a generally impermeable bed. Such spots bear no relationship whatsoever to the ancient or modern water courses (Fig. 49).

Diagenetically the gley forms from both hard (carbonatic) and soft (carbonate-free) waters. Siderites are precipitated in the carbonaceous strata. At the present time, gley forms in the taiga lakes of the northern USSR (carbonate-free) and in muds of the lakes on steppes (carbonatic environments). Its development in marine muds is generally arrested by the process of desulfurization, which leads to pyritization. According to Strakhov (1960), glauconite forms during the first stage of marine diagenesis. As the gley formation proceeds, leptochlorites next form and, as the Eh lowers, siderite, and finally pyrite form. The latter might replace all the other iron minerals.

As a result of this diagenesis, sands often lose all their mobile iron and take on a bleached appearance. The so-called "glass sands" may well be by-products of gley-forming processes.

Chapter 12

Epigenetic Processes in Reducing, Sulfidic Environments

THE SULFATO–SULFIDIC PROCESS

This process can be observed in the cementation zones of sulfide ore deposits (Saukov, 1951; Tatrinov, 1963). Sulfuric acid and sulfates of heavy metals which have descended from the oxidation zone above, react with primary sulfides. As a result, sulfides of heavy metals (mainly those of copper and silver) are formed. Hydrogen sulfide is generated locally. The waters no longer contain any oxygen, since this was consumed within the zone above. This process can be illustrated by the following equations:

$$2H_2SO_4 + CuFeS_2 \rightarrow CuSO_4 + FeSO_4 + H_2S$$

$$7CuSO_4 + 4H_2O + 4FeS_2 \rightarrow 4H_2SO_4 + 4FeSO_4 + 7CuS$$

$$Ag_2SO_4 + Cu_2S \rightarrow Cu_2SO_4 + Ag_2S$$

$$FeS_2 + ZnSO_4 \rightarrow FeSO_4 + S + ZnS.$$

The sequence of formation of the secondary sulfides is controlled in the cementation zone by the solubility products and the electrochemical properties of the sulfides. Considerable practical value is attached to these processes since they are instrumental in the formation of the secondary enrichment ores in many ore deposits.

An epigenetic zoning develops during the alteration of sulfide ore deposits as the sulfatic process of the oxidation zone is succeeded by the sulfidic in the cementation zone.

THE OXYSALT–SULFIDIC PROCESS

This process develops mostly in rocks in which the percolation of waters is difficult and waters stagnate. Microbiologic activity begins if the rocks contain even minute amounts of such organic substances as humus, petroleum, or bitumens. Oxysalts, particularly sulfates, are broken down by the microorganisms to obtain oxygen.

The oxidizing organic substances contribute carbon dioxide to the waters. The reduction of sulfates results in the formation of hydrogen sulfide. Gradually, the environment becomes strongly reducing. The Eh may drop as low as -0.5 V. Several species of sulfate-reducing organisms are known. Such bacteria thrive wherever organic matter decomposes in the presence of sulfates. A typical such desulfurization reaction is:

$$C_6H_{12}O_6 + 3CaSO_4 = 3CaCO_3 + 3H_2O + 3CO_2 + 3H_2S.$$

Desulfurization is common on the bottom of oceans, seas, salt lakes, and marshes. In the Black Sea, the process is active in deep trenches as well as in shallow bays. In general, the process is very active in oil fields and in bituminous rocks. The content of hydrogen sulfide in the waters of oil and gas fields may reach 2000 mg/liter (Shcherbakov, 1956).

Metals forming insoluble sulfides display very low migrational abilities during this process. Hydroxides and other compounds of ferric iron which might have been present in the rocks are reduced and converted to sulfides. Black colloidal sulfides such as hydrotroilite and melnikovite form. Their iron content is not great, but their coloring property is considerable.

Desulfurization is characteristic at the water–oil interphase, where it produces pyrite. Consequently, black sandstones containing both pyrite and bitumen are indicators of former oil and gas deposits. Gley impregnation on the periphery of the pyritization constitutes a zoning characteristic for oil and gas deposits. This will be discussed in greater detail in the next chapter.

Lead, zinc, copper, and other chalcophile elements are immobilized during this process. As a result, circulating ground waters are depleted in these elements. No secondary dispersion haloes are formed. An indiscriminate sampling of such waters fails to indicate hidden ore bodies.

Vigorous chemical oxidation of organic substances during desulfurization produces large quantities of carbon dioxide. This lowers the pH to about 6.5, according to A. K. Lisitsyn. Corrosion stylolites, and epigenetic calcite (Udodov et al., 1962).

The oxysalt–sulfidic process is common to both diagenesis and epigenesis. The diagenetic type predominated in oceans and salt lakes. It has been investigated by N. M. Strakhov.

A special variety of this process develops during marine transgressions, when marine waters spread over the sedimentary strata. The sea water, trapped in fissures and pores of the rock, becomes desulfurized. The pyrite which is forming imparts greenish- to bluish-gray color to the rocks, which, at a first glance, may appear to be impregnated with gley. However, this is not so, since pyrite and gley do not form simultaneously.

Together with Komarova and Kondratieva we have observed such a phenomenon in the basin of the Sumsar River in northern Fergana, where Alai limestones overlie red Bukhara alevrolites. The uppermost strata of these alevrolites (1 to 2 m thick) are green but contain stringers and inclusions of ferric hydroxides, which are the products of pyrite oxidation (Fig. 50).

When organic matter is present in amounts insufficient to reduce all the sulfates, some of these, among them gypsum, celestite, and mirabilite, will remain unaltered during this process despite the presence of hydrogen sulfide and sulfides. Also, the excess of hydrogen sulfide slows down the bacterial activity.

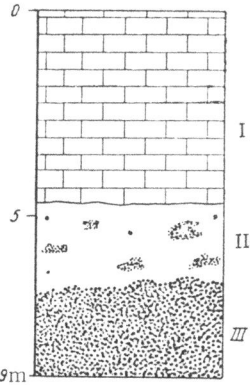

Fig. 50. The Saline–Sulfatic process in red rocks of the Bukhara layer (Paleogene), in northern Fergana. (I) Allai limestones; (II) green alevrolite (altered red rocks) with ferric hydroxides and pyrite; (III) red Bukhara alevrolites.

The oxysalt–sulfidic process develops in marshes of deserts and semideserts. It can also develop in humid climates if the rocks contain salt or are saturated with sea water. Such coastal marshes even exist in the tundra. Oxidation and gley formation may be associated with this process. In fact, a single paragenetic series or a more complex epigenetic zoning may develop. These phenomena are typical of coastal chlorsulfatic marshes or drying salt lakes.

Let us consider the Mulallah Lake in Uzbekistan. In spring and in autumn it is a typical marsh covered with a salt crust about 1 cm thick. The underside of this crust supports algae, and is therefore greenish. The brine or the wet gypsiferous sand lie beneath it. It is obvious that an oxidizing environment develops immediately under the salt crust, since oxygen is exuded by the algae. This accounts for the oxidation of the upper 3 to 5 cm of sand which is brown in color. Below this portion, the sand turns black due to its content of hydrotroilite, and it exudes hydrogen sulfide.

The zone affected by this oxysalt–sulfidic process is only 3 to 5 cm thick. The organic matter of the algae, dead or living, serves to induce the desulfurization. Below this zone the sand is bluish gray, due to the formation of gley which begins when the supply of hydrogen sulfide is depleted. This is a typical epigenetic zoning: oxidized layer → hydrotroilitic layer → gley layer. Since the intensity of algal growth fluctuates with the time of the year, the oxysalt-sulfidic process ebbs, and its individual zones are not permanent.

This sample proves that the activity of the living matter and the biologic cycling of atoms induce the coexistence in the same space of strongly oxidizing and strongly reducing environments. The oxysalt–sulfidic process can be recognized by means of the following criteria:

1. The presence of hydrogen sulfide in sediments, rocks, or waters.
2. The black color of rocks and sediments caused by hydrotroilite and other iron sulfides.
3. The sulfatic content of the water.
4. The association of hydrotroilite with gypsum or other sulfates.

The criteria indicating the absence of this process are:

1. Brown and red colors of ferric hydroxides.
2. The presence of ocherous and gley-impregnated spots, which indicate the migration of iron.

During regional uplift and the development of the topographic relief, slime sediments of lakes and marshes might be moved into the eluvial environment.   Such an eluvium is recognized by its considerable gypsum content. Its salt content, as a rule, is negligible.   Salts are leached out from the upper 2 or 3 m. In other words, the upper layers of this weathered crust have lost their soluble content and become a species of gypsum capping.

Such cappings are usually colored reddish for the first 0.5 m from the surface due to films of ferric hydroxide. It is only logical to expect in this connection that when ground waters were close to the surface during the formation of the salt, hydrotroilite formed as a result of desulfurization. After the erosion base was lowered and the sediments dried, weathering and oxidation set in. Hydrotroilite is not stable under such conditions, and was converted into films of ferric hydroxide.

We have often seen such phenomena in the eluvium of Tertiary and Quaternary sediments in Turkmenia and western Uzbekistan.

Thus, the criteria for the recognition of the former activity of this process are:

1. Gypsum and water soluble salts.
2. Reddish to brownish coloring of gypsum imposed by the oxidation of hydrotroilite.

In past geologic epochs, the process has been common in many arid regions of the USSR. During the Neogene and Pleistocene periods, the ground water table was close to the surface on the Central Asiatic plains. The waters were sulfatic and contaminated the soils. The modern eluvium still contains relict salt and gypsum in marshes (Perel'man, 1959).

THE SODA–HYDROGEN SULFIDE PROCESS

This process takes place in marsh soils and bituminous rocks, particularly in petroliferous deposits undergoing destruction.

Percolating through such rocks, waters quickly lose their oxygen content by oxidizing the organic matter of the rocks. In turn, this spurs a proliferation of microbiologic activity and desulfurization.

As a result, the composition of the waters changes. Having previously carried sodium sulfate, these waters now begin to carry carbonate and bicarbonate of sodium.* Their pH increases to as much as 11, since hydrogen sulfide and carbon dioxide accumulate:

$$Na_2SO_4 + 2C_{org} + 2H_2O \rightarrow 2NaHCO_3 + H_2S.$$

Under such conditions, the waters become similar to waters of the soda catagenesis of sandstones and yield the same epigenetic products. In the course of this reaction the rocks become bleached and can be seen from afar. However, a secondary deposition of migrating humus might camouflage this bleaching. Darkening might also be caused by the formation of pyrite. The desulfurization and rise in pH following the contact of water with petroleum results in the precipitation of calcite in the carbonaceous collectors. The precipitation effectively "seals" the oil deposit from the circulating ground waters. Such changes are best seen in bituminous limestones.

Analogous processes develop in marsh soils where we find many indications of the soda process, including the migration of humates, silication, the pseudomorphism of calcite after gypsum, and the formation of secondary aluminosilicates. However, in this case the formation of soda is related to the oxidation of vegetal remnants and humates but not bitumens.

This soda–hydrogen sulfide process apparently operates in salt lakes and swamps. The Eh of the brine reaches 500 mV and the pH can increase to 10.48 (at Searles Lake in California [Becking et al., 1963]). These waters also form in coal veins, e.g., on Donbass or Kuzbass. These have not received sufficient study.

---

*Depending on the chemical composition of the waters and the rocks, the biologic reduction of sulfates may also produce waters enriched in chlorides and calcium carbonate.

Chapter 13

Geochemical Barriers

Lithologic boundaries in the supergene zone where conditions of migration change drastically and concentrations of chemical elements begin to rise are called geochemical barriers. *

A drop in the migrational ability at the barrier results in the formation of mineral ore bodies. The high metal contents of ores can be largely explained by their precipitation at the time and place of some drastic change in environmental conditions. Any gradual change in some environmental conditions would bring about a gradual increase or decrease in metal content. Low-grade ores and barren stretches may be best explained in this way. Geochemical barriers may be classed as syngenetic, diagenetic, and epigenetic. Here we shall discuss only the epigenetic barriers.

On maps, the geochemical barriers are best represented by lines of demarcation of the fields of influence of one or another process. This, in turn, shows the geochemical barriers to be integral parts of epigenetic zoning.

In soils, the barriers coincide with the boundaries between individual soil layers. Barriers are also encountered in weathered crusts and in water-bearing strata. Their dimensions vary. Macrobarriers delimit the entire ore bodies in a given ore deposit, while microbarriers isolate smaller units, such as soil layers.

Many economic mineral deposits are localized on geochemical barriers, the intercalation of strata of different permeabilities is particularly notable in this respect. The ore bodies are confined to strata which carried waters. The impermeable intercalations remain barren.

*In one form or another, the concept of geochemical barriers already exists in some geological sciences. For example, "chemical contradiction" (Pustovalov, 1940) in sedimentary petrography represents such a barrier.

THE PRINCIPAL TYPES OF GEOCHEMICAL BARRIERS

The concentration of chemical elements at the barrier is caused by changes in pressure or in some other physicochemical factor. The barriers vary selectively depending on the modes of migration or concentration. These types may be distinguished: mechanical, physicochemical, and biologic.

Mechanical barriers develop in places where the velocity of water or air drastically changes. Such barriers are, therefore, characteristic of sedimentation. They account for the formation of placers of gold, tin, zirconium, titanium, etc. Mechanical barriers also control the configurations of mechanical haloes formed about ore deposits. Such phenomena have not been adequately studied. Solovov (1959) advocated the use of mathematical analysis for their investigation.

Physicochemical barriers are more varied but are also better known. They develop due to changes in factors such as redox conditions. The principal types are:

1. The oxygen type for Fe, Mn, Co, and S.
2. The hydrogen sulfide, reducing type for V, Fe, Cu, Co, As, Se, Ag, Ni, Zn, Cd, Hg, Pb, and U.
3. The sulfatic and carbonatic type for Ca, Sr, and Ba.
4. The alkaline type for Ca, Mg, Sr, V, Cr, Mn, Fe, Co, Ni, Cu, Zn, Cd, and Pb.
5. The acid type for SiO_2.
6. The evaporate type for Li, N, F, Na, Mg, S, Cl, K, Ca, Zn, Sr, Rb, Mo, I, and U.
7. The sorptive type for Mg, P, S, K, Ca, V, Cr, Co, Ni, Cu, Rb, Mo, Zn, As, Hg, Pb, Ra, and U.

The oxygen barrier develops where a reducing environment is suddenly succeeded by an oxidizing one.

According to Germanov (1955), oxygenated ground waters can penetrate to considerable depths below the ground water table. In one mountainous region, such oxygenated waters were identified at 600 m below the erosional base. In another region, oxygenated waters were discovered in sandstones at a depth of 626 to 644 m.

According to Babinets (1961), oxygenated waters penetrated down to 250 m along faults in the gneissic complex on the Ukrainian Shield near the town of Khmelnik. In the Krivorozh e bolt, products of ancient exogene oxidation were encountered at a depth of 1000 m.

These observations lead to the conclusion that an oxygen barrier was in existence at a still greater depth.

The interphase between gley water and oxygenated water (or air) is another oxygen barrier. The hydroxides of iron and manganese epigenetically deposit in such places. This may well be the origin of the red coloring of rocks along fault planes where deep-seated waters come up to the surface. Gley waters emanating from gas or oil structures or from some deep-seated igneous body become oxidized on contact with air. It is possible that micro-organisms oxidize bivalent iron and manganese. It has been established that iron bacteria oxidize ferrous iron to ferric. Manganese should be oxidizable in the same way.

Deposits of native sulfur apparently form at the oxygen barrier, where hydrogen sulfide is oxidized. The oxidizing agent is air. Dissolved in water, it penetrates to considerable depths during tectonism. Hydrogen sulfide might be the product of bacterial activity. Migrating upwards, hydrogen sulfide oxidizes and sulfur precipitates in the elemental state:

$$2H_2S + O_2 \rightarrow 2H_2O + 2S.$$

Such was the probable origin of sulfur in the Shor-Suh sulfur deposit in Fergana. Sulfate-reducing bacteria are known to exist in the nearby oil fields. They produce hydrogen sulfide, which is then oxidized by another group of bacteria (Fig. 51).

Hydrogen sulfide can also oxidize without the aid of bacteria. In such cases, sulfur deposits are formed by waters ascending from the underlying oil pools.* The Gaurdak and other sulfur deposits of southeastern Turkmenia probably originated in this manner. There, ground waters deposited sulfur in solution cavities. Gypsum, calcite, and fluorite formed simultaneously (Petrov, 1955).

In the western Ukraine, sulfur was precipitated in places where oxygenated ground waters mingled with those carrying hydrogen sulfide (Goleva, 1959).

It is possible that sulfur bacteria took part in the precipitation of sulfur. According to Goleva, the time required for the formation of the Ukrainian sulfur deposits was, from the geologic point of view, relatively short. The epigenetic sulfur deposits in caprocks of salt domes on the Gulf Coast (USA) were apparently formed

---

*This process is diametrically opposed to those forming epigenetic uranium deposits. The latter are formed by waters descending into the oil deposits.

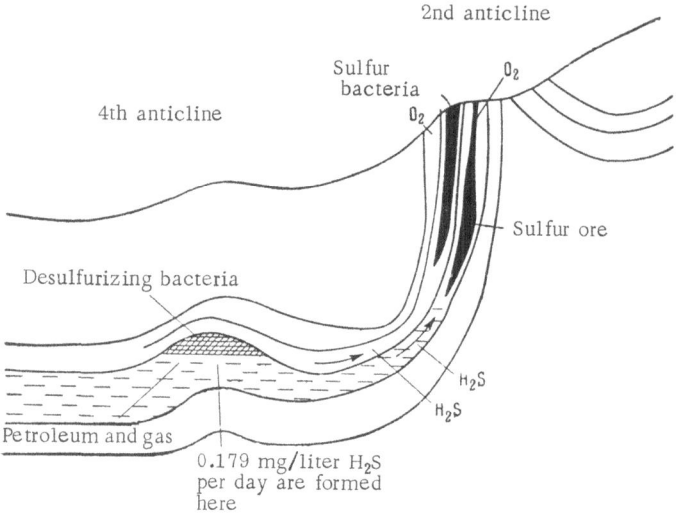

Fig. 51. The participation by different groups of microorganisms in the formation and destruction of the Shor-Suh sulfur deposit (revised after Kuznetsov et al., 1963).

by the action of microorganisms at the oxygen barrier (Kuznetsov et al., 1963).

When the hypogene nonsulfatic brines of Ca, Sr, and Ba mingle with oxygenated sulfatic waters which have infiltrated from above, secondary gypsum, celestites, and barites precipitate. Silica may also precipitate. Hydrogen sulfide will precipitate any iron, zinc, or other metal which may be present in the ascending waters. Evolving during the desulfurization, carbon dioxide helps dissolve the limestone, which is metasomatically replaced with native sulfur (Sokolov, 1959).

The reducing, H_2S type of barrier develops in aquifers when either oxidizing or weakly reducing waters encounter hydrogen sulfide. The H_2S may be the product of rotting organic matter, or it may have arrived in the form of gas or in solution. It precipitates heavy metals in the form of insoluble sulfides.

This mechanism is instrumental to the formation of copper and uranium ores, and, possibly those of lead, zinc, and some other metals. Pyrite and other sulfides occurring in the rock indicate that hydrogen sulfide was evolving at one time.

A variety of this barrier develops at the contact of water with oil when oxygenated waters invade petroliferous or bituminous rocks. The process is also important in the formation of epigenetic uranium deposits. Soviet geologists have published numerous papers concerning this subject (Germanov, 1960, 1961; Kholodov et al., 1961; Evseeva and Perel'man, 1962).

It is a well-known fact that uranium possesses a high migrational ability, its content in waters reaching up to n · 10^{-4} g/liter. On entering petroliferous or bituminous rocks, these waters encounter a reducing barrier. Any free oxygen which could have been dissolved in water was used up in the oxidation of organic substances, so that any further oxidation will be at the expense of sulfates, vanadates, nitrates, ferric hydroxides, etc.

The activity of the sulfate-reducing bacteria is pronounced. Its by-products — CO_2 and H_2S — accumulate in the water. Calcium carbonate dissolves in places of CO_2-evolution and leaves behind cavities and stylolites. Secondary calcite reprecipitates elsewhere. The precipitation of silica causes the silication of limestones. Because of the destruction of organic matter, the limestones become bleached.

A reducing environment, which is created in the water near its contact with oil, is conditioned by the hydrogen sulfide which is generating. In turn, it conditions the pyritization of the rock and the fixation of tetravalent uranium (as pitchblendes or other uranium "blacks").

Kholodov has worked out the zoning typical of ore deposition in bituminous limestones (Fig. 52). The epigenetic zones reflect a succession of events taking place when oxygenated waters penetrated into the limestones containing the reducing agents:

I. Oxidation zone. This zone is composed of porous limonitic limestone which usually runs 8 · 10^{-5} to 2 · 10^{-4} % uranium and n · 10^{-4} % vanadium. Its content of organic carbon is generally less than 9.01%.

II. Zone of bleached and pyritized limestones. This is a zone of "biochemical oxidation." Having been bleached, it contains very little organic matter. The uranium content is under 8 · 10^{-4} % and the vanadium content is n · 10^{-4} %. Iron produces epigenetic pyrite and marcasite. An intensive redistribution of calcite is characteristic.

III. Ore zone. Limestones comprising this zone contain black uraniferous smudges and pitchblendes in association with pyrite and bitumens. These ores are noted for their somewhat higher values of vanadium, molybdenum, nickel, and cobalt. The silica-

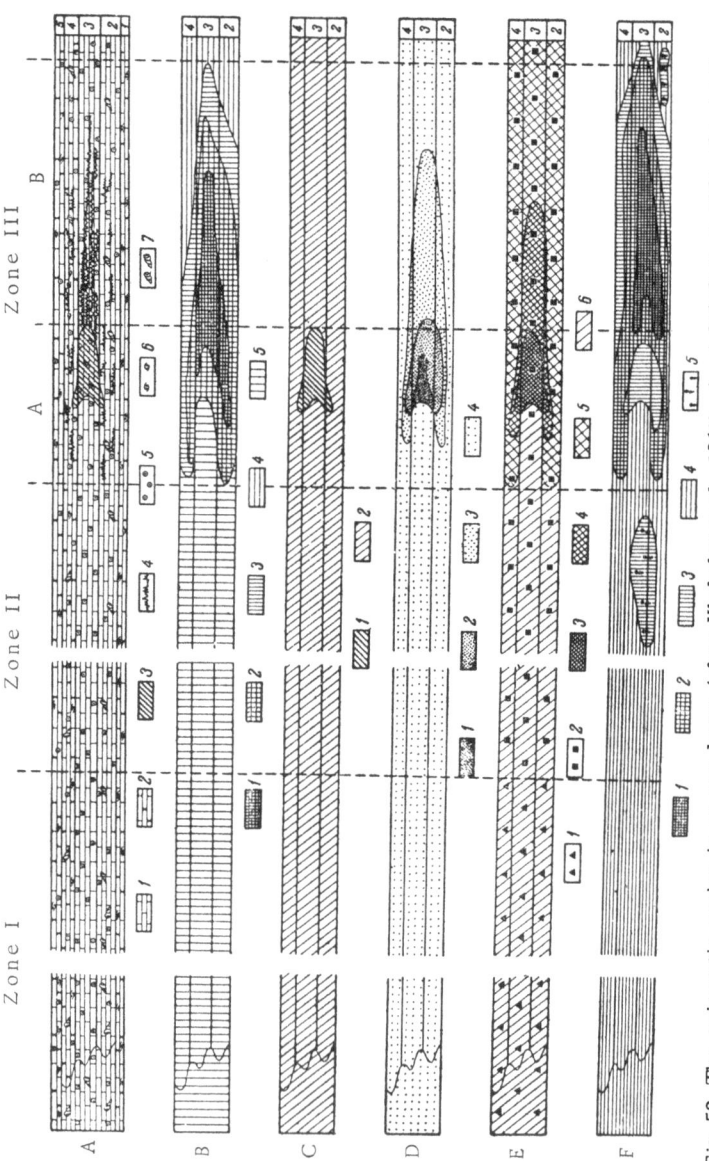

Fig. 52. The epigenetic zoning in an ore layer (after Kholodov et al., 1961). A. A lithologic profile, showing the facies: (1) limestones; (2) dolostones; (3) silica concretions; (4) stylolitic seams; (5) oolites; (6) pelecypods; (7) gastropods. B. Distribution of uranium, %: (1) to (5) various uranium concentrations in decreasing order. C. Silica, %: (1) 50 to 25; (2) traces. D. Vanadium, %: (1) 0.n; (2) 0.0n; (3) 0.00n; (4) 0.000n. E. Total iron, %: (1) region of hydroxides of Fe; (2) region of sulfides of Fe; (3) over 0.60; (4) 0.60 to 0.40; (5) 0.40 to 0.25; (6) 0.25 to 0.10. F. Organic carbon, %: (1) 0.60 to 0.40; (2) 0.40 to 0.20; (3) 0.20 to 0.10; (4) under 0.10; (5) blocks saturated with liquid petroleum.

tion of limestones is typically in the form of concretions (the subzone IIIa) or stylolitic sutures.

IV. Zone of unaltered or slightly altered bituminous limestones. The rocks in this zone are typically gray and contain disseminated pyrite. Their uranium content is normal.

The first zone develops where oxygenated waters penetrate into oil formation and oxidize organic matter and pyrite. As the free oxygen contained in these waters is spent, the biochemical oxidation of the petroleum components takes over. Carbon dioxide and hydrogen sulfide accumulate in the water and its pH lowers. A reducing barrier is thus created. Pyrite again begins to precipitate. Reprecipitating calcite forms stylolites. Silica precipitates wherever the waters become acid enough.

Uraniferous ores are not confined to the barrier itself but keep forming below it. The explanation for this is that while desulfurization took place in the second zone, the pH was still high enough to permit the existence of soluble carbonate complexes of uranium. In the third zone, the pH drops to 6.4 and the complexes are destroyed.

Soviet geologists have studied still another type of uranium deposit which is confined to the reducing barrier. Such deposits lie on the periphery of artesian basins and are formed by descend-

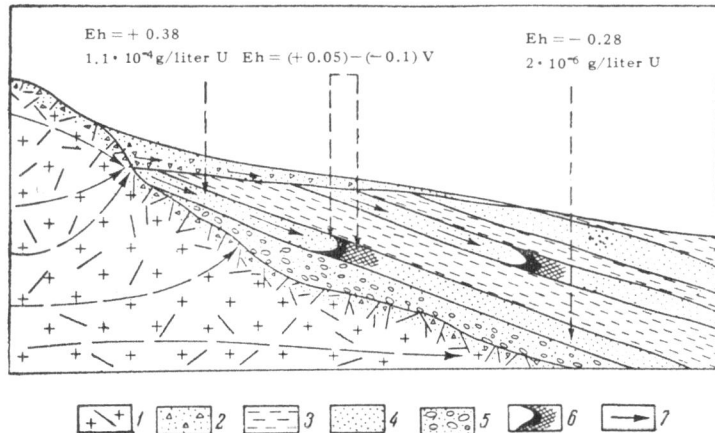

Fig. 53. The localization of uranium ore in sedimentary strata (after Evseeva, 1962). (1) Granites; (2) deluvial–proluvial sediments; (3) impermeable argillaceous rocks; (4) sandy water-bearing strata; (5) conglomerates; (6) ore bodies; (7) the direction of the water flow.

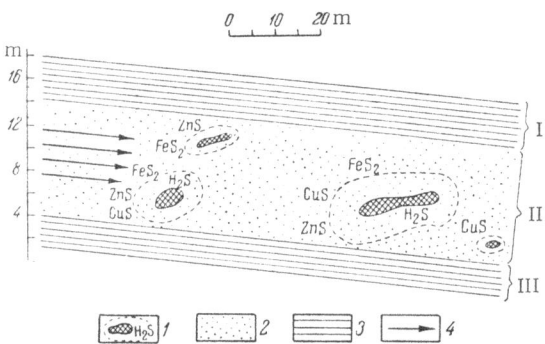

Fig. 54. The form ation of ore of the type of cupriferous sandstones. Decaying vegetal remnants are frequently present in the aquifers (the Permian rocks of the Ural region). (1) Decaying organic matter surrounded by the halo of H_2S; (2) sands; (3) red-brown alevrolite; (4) the direction of ground waters carrying bivalent Fe, Cu, and Zn. (I) and (III) Impermeable horizons; (II) an aquifer with weakly oxidizing to weakly reducing pH.

ing oxygenated waters. The ore is restricted to strata of gray unconsolidated, permeable sandstones. The oxygenated ground waters contain $n \cdot 10^{-5}$ to $1 \cdot 10^{-4}$ g/liter uranium, and have Eh values over 0.25 V. These waters oxidize the gray sandstones, turning them brown due to limonitization. In such cases a local oxidation zone develops. Below it, sandstones resume their gray color, the uranium content drops to 1 to $3 \cdot 10^{-6}$ g/liter and the Eh becomes negative (-0.05 to -2 V).

The "manto" ore bodies were formed by the precipitation of uranium from oxygenated waters at the barrier where the oxidation zone ended. This barrier is easily recognized by the change from the limonitized to gray pyritic sandstones (Fig. 53).

The formation of copper ores in red rocks is controlled within the catagenetic subzone by a reducing barrier. In the absence of hydrogen sulfide, copper possesses a fairly high migrational ability in both oxidizing and reducing environments. Various epigenetic processes, including sulfatic, acid, sodic, and gley processes, are therefore, favorable for its migration.

Hydrogen sulfide precipitates copper in the form of insoluble sulfides. A reducing H_2S barrier also develops in the vicinity of decaying organic matter, and copper segregates in sandstones about vegetal remnants.

Two types of copper mineralization are recognized in "cupriferous" sandstones (Perel'man and Borisenko, 1962). The first of these two types is due to the precipitation of copper from ground waters in places of local reducing environments, such as decaying organic matter. Since the red rocks were oxidized, they contain very little organic matter, the accumulations of which are small and disconnected.

As a result, a multitude of small "reducing centers" appears in the catagenetic subzone. Each one precipitates copper. Some 4000 such centers have been counted in the Ural region. Several hundred "centers" are known in the Devonian red rocks of Kazakhstan and in the Permian red rocks of the Donbass. This type of copper mineralization has no practical value.

As repeatedly stated, hydrogen sulfide is generated during the desulfurization of sulfatic waters emanating from oil and gas pools. Migrating in waters, it precipitates most heavy metals. This accounts for the second type of copper ores in cupriferous sandstones. The copper deposits of Dzhezkazgan are examples of this (Germanov, 1961; Perel'man and Borisenko, 1962).

We believe that although the host rocks were formed in oxidizing conditions, the development of the superimposed stable and long-lasting barrier has determined the size and character of the ore bodies.

In the case of ore bodies of the first type, the source of the reducing medium was syngenetic — the red rocks themselves. In the second case, it was epigenetic — oil, gas, bituminous rocks, etc. The mineralization was epigenetic in both cases. The genesis of ore deposits of the second type is shown in a simplified way in Fig. 55.

In the epoch which preceded the ore making, the dry land consisted of copper-bearing rocks. Secondary dispersion haloes existed about the ore deposits and particularly about their outcrops. Their erosion and the redeposition of the eroded material led to the sorption of copper by the red rock sediments. Subsequently, these red rock sediments were intensely worked over by ground waters. This is substantiated by the existence of gley-impregnated strata and by numerous indications of soda catagenesis. To start with, the waters could have been weakly mineralized and lacked hydrogen sulfide, or there would have been no migration of iron. The gray and green sandstones were thus formed. The waters were copper-bearing, having leached Cu out of the enclosing rocks. These waters could have carried hydrogen sulfide into the lower strata of the suite of red rocks after obtaining it from the bituminous rocks, oil, or gas. If so, a vertical hydrochemical zoning could

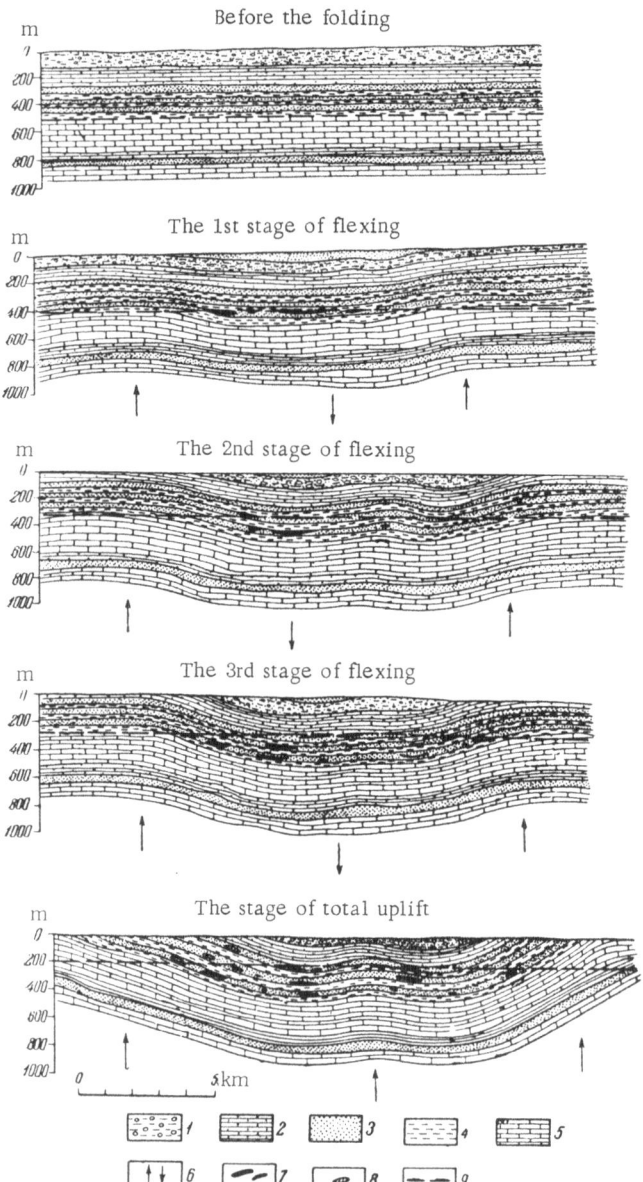

Fig. 55. The formation of ore of the type of cupriferous sandstones of the second type. (1) Conglomerates, sands, clays; (2) clayey alevrolites; (3) sands, sandstones, sandy alevrolites; (4) clays, argillites, clayey argillites; (5) bituminous limestones; (6) the direction of tectonic movements; (7) ore bodies and ore ozones; (8) oxidized ores; (9) geochemical barrier. (The direction of motion of cupriferous solutions is perpendicular to the plane of this drawing.)

have developed in the red rocks during the mineralization epoch. It would have consisted of the following zones:

1. The zone of oxygenated waters with mobile copper and im-mobile iron.
2. The zone of gley waters with no oxygen but with mobile copper.
3. The zone of H_2S waters with the immobilized iron and copper.

The reducing, H_2S barrier developed on the boundary between the second and third zones and precipitated copper and its satel-lites. Very little if any pyrite formed because no iron could be leached in the carbonate environment of the second zone.

When the region began to subside, the strata sank gradually into the reducing environment. Their mineralization acquired a layered aspect. The absence of ore at the bottom of the suite is probably due to those strata remaining always within the third zone of hydrogen sulfide. Thus, the waters descending to that depth would already be depleted of their copper content. Somewhat similar views were expressed by Germanov (1961).[*] There are two possible cases of the precipitation of metals at the reducing, H_2S barrier: with excess or with insufficient hydrogen sulfide.

Should the supply of H_2S be ample, the precipitation of all dis-solved metals will occur and their insoluble sulfides will be formed. Such ores or mineralized rocks are polymetallic. Again there are two additional possibilities: (1) if the ores contained sulfides of several metals, then the supply of hydrogen sulfide must have been in excess; and (2) if the ore is monometallic, or, contained only two or three metals at the most, either the supply of H_2S was insufficient or the solution was monometallic.

Let us visualize a flow of ground water carrying various heavy metals in solution and moving through an aquifer toward a reducing H_2S barrier. Let the concentrations of S^{2-} or HS^- be insufficient to precipitate all the active concentrations of the metals.

Under such conditions, the precipitation of sulfides will be controlled by the principle of retardation. If the concentration of

[*]Here we are exploring a hypothesis that needed verification. We do not intend to re-view all possible hypotheses on the origin of copper ore deposits. Suffice it to say that there are also sedimentary and hydrothermal theories in regard to the Dzhezkazgan copper ore.

Table 31. The Metallic Content of Waters Entering the H_2S Barrier, the Isobar Potentials of the Reactions of Sulfide Formation (ΔZ), the Standard Isobar Potentials ($\Delta Z°$) and the Solubility Products at 25°C and 1 atm

Metallic cation	Amount in the water entering the barrier		Characteristic of the reaction: $Me + S^{2-} \rightleftharpoons MeS$		Solubility product of MeS for $\Delta Z°$
	mole/liter	g/liter	ΔZ	$\Delta Z°$	
Ag^+	$1 \cdot 10^{-7}$	$1.07 \cdot 10^{-5}$	-35.85	-68.58	$1 \cdot 10^{-50.4}$
Cu^{2+}	$1 \cdot 10^{-6}$	$6.3 \cdot 10^{-5}$	-27.50	-49.33	$1 \cdot 10^{-36.2}$
Cd^{2+}	$1 \cdot 10^{-7}$	$1.12 \cdot 10^{-5}$	-13.93	-37.12	$1 \cdot 10^{-27.3}$
Pb^{2+}	$1 \cdot 10^{-7}$	$2.07 \cdot 10^{-5}$	-13.25	-36.44	$1 \cdot 10^{-26.8}$
Sn^{2+}	$1 \cdot 10^{-8}$	$1.18 \cdot 10^{-6}$	-11	-35.53	$1 \cdot 10^{-26}$
Zn^{2+}	$1 \cdot 10^{-6}$	$6.5 \cdot 10^{-5}$	-12.49	-34.32	$1 \cdot 10^{-25.2}$
Co^{2+}	$1 \cdot 10^{-7}$	$5.9 \cdot 10^{-6}$	-5.91	-29.1	$1 \cdot 10^{-21.4}$
Ni^{2+}	$1 \cdot 10^{-7}$	$5.9 \cdot 10^{-6}$	-5.08	-28.27	$1 \cdot 10^{-20.8}$
Fe^{2+}	$1 \cdot 10^{-5}$	$5.6 \cdot 10^{-4}$	-4.72	-25.12	$1 \cdot 10^{-18.4}$
Mn^{2+}	$1 \cdot 10^{-6}$	$5.5 \cdot 10^{-5}$	$+4.23$	-17.6	$1 \cdot 10^{-12.9}$
Cu^+	—	—	—	-66.7	$1 \cdot 10^{-49.04}$

the sulfide ion were 10^{-10} mole/liter ($3.2 \cdot 10^{-9}$ g/liter), the reactions would be:

$$Cu^{2+} + S^{2-} \rightleftharpoons CuS,$$

$$Pb^{2+} + S^{2-} \rightleftharpoons PbS \text{ etc.}$$

Let the metal content of the waters arriving at the barrier be equal to the "background content," i.e., between $1 \cdot 10^{-6}$ and $1 \cdot 10^{-8}$ mole/liter ($n \cdot 10^{-5}$ to $n \cdot 10^{-6}$ g/liter).

Then, the values of ΔZ and ΔZ^0 are calculated as shown in Table 31.*

*The aliquot concentrations of metals cannot be calculated precisely from the solubility products. Cations like Fe^{2+} are used in calculations of the solubility products, although they are rarely present in natural waters. As a rule, natural waters carry HS^- and not S^{2-}. Thus, the actual solubility products are seldom known. Judging by the experiments of Olshansky and others, the actual solubility products should be greater than those commonly used (Wolfson, 1962).

Nevertheless, a very small quantity of metals does occur in the form of simple cations. On the basis of the exceedingly small solubility products of sulfides we can assume that the concentrations of these cations should also be small and insufficient for the precipitation.

The solubility products of the most important sulfides are given in Table 31. Obviously, the variations between individual values are considerable. We therefore assume that the solubility products for the hydrosulfides are proportional to the solubility products of the corresponding sulfides. For example, the solubility product for $Cu(HS)_2$ should be smaller than the solubility product for $Zn(HS)_2$. Since the final products of the reactions are sulfides, we are justified in using the solubility products of sulfides for the purposes of illustration.

The following example illustrates the method of calculating the sulfide precipitation:

$$Zn^{2+} + S^{2-} \rightleftharpoons ZnS.$$

The standard isobar potential is calculated by using the data of Tables 15 and 16 to be -34.32 for the reaction of ZnS precipitation. The corresponding isobar potential is calculated in the following manner:

$$\Delta Z = RT \ln \frac{a_{ZnS}}{a_{Zn^{2+}} a_{S^{2-}}} + \Delta Z^0 = 1.364 \quad \log \frac{a_{ZnS}}{a_{Zn^{2+}} a_{S2-}} + (-34.32).$$

But the condition was that $a_{ZnS} = 1$. Hence

$$a_{Zn^{2+}} = 1 \cdot 10^{-6}; \quad a_{S^{2-}} = 1 \cdot 10^{-10}.$$

Then

$$\Delta Z = 1.364 \log \frac{1}{10^{-6} \cdot 10^{-10}} - 34.32 = -12.49.$$

Isobar potentials for all other reactions are calculated in the same manner.

A simple calculation shows that if the precipitant, S^{2-}, remains the same, a change of one order in the molar concentration of the metal in water modifies the ΔZ of the sulfide precipitation by 1.364. Therefore, the sequence in which metals with considerably different isobar potentials precipitate does not depend on the initial concentrations of those metals. For example, copper always precipitates first from H_2S-deficient solutions regardless of the concentrations of Cu and Zn.

However, if the isobar potentials are nearly the same, the sequence of their precipitation does depend on the initial concentrations. As an example, we consider the pairs Pb^{2+} and Zn^{2+} or Ni^{2+} and Co^{2+}. It follows from the data of Table 31 that silver and copper should precipitate first under such conditions. The precipitation of Cd, Pb, Sn, and Zn should precede the precipitation of Fe, Ni, and Co, although not necessarily in the given order. Actual sequences vary with the concentrations. Lead, particularly, tends to precipitate either earlier or later than zinc. Considering, however, that zinc is usually more abundant, its sulfide, ZnS, should precipitate ahead of PbS. The same holds for the precipitation of NiS and CoS: nickel should precipitate ahead of cobalt.

Kengir anticline

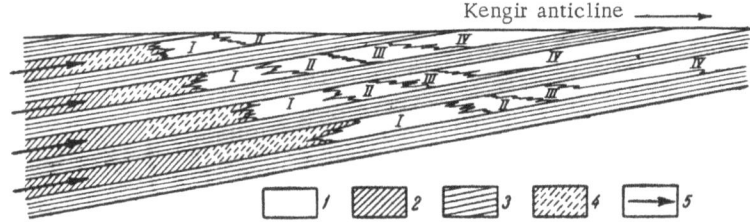

Fig. 56. Distribution of ore bodies and mineralized zones in a Dzhezkazgan ore deposit [after Narkelyun (1963), revised]. (1) Gray mineralized sandstones and red rocks; (2) sandstones; (3) elevrolites and argillites; (4) reddish-gray and grayish-red sandstones; (5) the probable direction of ore solutions during the epigenetic stage. The mineralized zones: (I) Chalcocitic; (II) bornitic; (III) chalcopyritic; (IV) pyritic.

The isobar potential of the MnS precipitation is positive. Hence, manganese does not precipitate under such conditions ($Mn^{2+} = 1 \cdot 10^{-4}$ mole/liter or $5.5 \cdot 10^{-3}$ g/liter). To start precipitating, its concentration should rise to $1 \cdot 10^{-2}$ mole/liter (0.55 g/liter). This rise corresponds to the change in the isobar potential from +1.51 to −1.21. Thus, alabandite, MnS, forms only at high concentrations of either Mn or H_2S. In media deficient in H_2S, there will be no alabandite formed.

The ΔZ and $\Delta Z°$ values for copper differ radically from the same functions for other metals. This might be the determining factor in the formation of copper ores in the red rocks of the Urals, Kazakhstan, and Central Asia. Hydrogen sulfide apparently was insufficient to precipitate other metals if other metals were available at the time.

Copper sulfides predominate in ore deposits of the cupriferous sandstone type, while pyrite is present in negligible amounts. The scarcity of pyrite is not due to a lack of iron but is due to H_2S deficiency. The latter was only sufficient to produce chalcocite and covellite. The absence of cobalt and nickel does not indicate that these metals were absent in the mineralizing solutions. It only proves that the H_2S supply was inadequate.

An epigenetic zoning is usually observed in the distribution of ore sulfides. Copper sulfides precipitate closer to the barrier on the side of oxygenated waters when H_2S is deficient. Lead and zinc localize further within the H_2S zone. In keeping with the principle of retardation, silver in Dzhezkazgan occurs mainly in bornite and tetrahedrite and not in galena. According to Kuganskaya, the Dzhezkazgan chalcocite runs up to 1% Ag, and galena 0.0n% Ag.

The following list gives the mineralogical zoning of Dzhezkazgan ores as worked out by Narkelyun (1963). To further clarify it, we have added data on the standard isobar potentials and solubility products (Fig. 56).

Mineral zones	$\Delta Z°$	Solubility products
Chalcocite, Cu_2S	-66.7	$10^{-49.4}$
Bornite, Cu_5FeS_4	-	-
Chalcopyrite, $CuFeS_2$	-	-
Galena, PbS	-36.44	$10^{-26.8}$
Sphalerite, ZnS	-34.32	$10^{-25.2}$
Pyrite, FeS_2 (values given are for FeS)	-25.12	$10^{-18.4}$

The exact order of succession is controlled by the principle of retardation, but in any event the transitions from one zone to another are gradational.

Thus, the fractional precipitation from a solution containing copper, lead, and other metals cannot be satisfactorily explained by the mere fact that lead requires a stronger reducing environment than copper. The key factor is the quantity of hydrogen sulfide available for the precipitation of metals.*

The principle of retardation explains the occurrence of chalcocitic ores in reddish-gray sandstones (a reducing medium) and of chalcopyritic ores in gray rocks (a stabilized reducing medium, enriched in H_2S). A similarity in $\Delta Z°$ values for Cu_2S and Ag_2S explains the greater affinity of silver toward bornite–chalcocite ores than toward galena–sphalerite ores. According to Domarev, who quoted Narkelyun, a similar zoning is encountered in Mangyshlak cupriferous sandstones.

While the principle of retardation does work satisfactorily in connection with cupriferous sandstones, it does not control the precipitation of all sulfides. At this stage of our investigations, the primary merit of the principle is for providing a new method of analysis of natural processes.

Sulfatic and carbonatic barriers develop where waters of this type mingle with waters of some other type which carry Ca, Sr, and Ba. The mingling of hypogene chloridic brines with infiltrating waters carrying sodium sulfate or sodium bicarbonate is an exam-

---

*Since the valence of Cu, Pb, or Zn does not change during the precipitation of their sulfides, the precipitation is not affected by the value of Eh as long as H_2S is present.

ple. Such phenomena take place on a down-warped wing of an artesian basin, along which the hypogene waters rise to the surface.
There, the typical exchange reactions are:

$$
\left.
\begin{array}{l}
Ca^{2+} \\
Sr^{2+} \\
Ba^{2+}
\end{array}
\right\}
+ \; SO_4^{2-} \longrightarrow
\left\{
\begin{array}{l}
CaSO_4 \; \text{(Anhydrite, etc.)} \\
SrSO_4 \; \text{(Celestite)} \\
BaSO_4 \; \text{(Barite)}
\end{array}
\right.
$$

Chloridic   Infiltrating   Precipitates
  brines   waters

$$
\left.
\begin{array}{l}
Ca^{2+} \\
Sr^{2+} \\
Ba^{2+}
\end{array}
\right\}
+ \; CO_3^{2-} \longrightarrow
\left\{
\begin{array}{l}
CaCO_3 \; \text{(Aragonite, calcite)} \\
SrCO_3 \; \text{(Strontianite)} \\
BaCO_3 \; \text{(Witherite)}
\end{array}
\right.
$$

Chloridic   Infiltrating   Precipitates
  brines   waters

Tectonic uplifts are probably the commonest causes of such
mingling.

An alkaline barrier develops where acid waters are suddenly
neutralized or even made weakly alkaline. Waters draining from
sulfide ore deposits or from ultrabasic igneous rocks are generally
acid. They carry large amounts of metals in solution. When these
waters come in contact with limestones, their pH is suddenly
changed. Metals begin to precipitate, filling the solution cavities
or replacing the limestones metasomatically.

Such was the origin of nickel ore deposits of the Ufalei type in
the central Urals. Acid waters emanating from a massif of ser-
pentinite carried significant amounts of nickel. They changed the
pH and precipitated nickel ore on coming into contact with limestones. Iron ore deposits of the Urals, precipitated in limestones,
are another example.

Alkaline barriers typically develop in soils, weathered crusts,
and the dispersion haloes around sulfidic ore deposits.

A weakly acid environment is generally developed in the upper
layers of steppe soils. Its waters, percolating down, encounter an
illuvial carbonatic layer at a relatively shallow depth (0.5-1.0 m).
An alkaline barrier develops at the contact and the conditions become favorable for the precipitation of a number of metals.

The acid barrier is the least common in the supergene zone,
because any decrease in the pH increases the mobility of chemical
elements and their compounds. The only exception to this is silica,

which is poorly soluble in acid waters. Rocks become silicated when neutral or alkaline waters encounter an acid barrier and dump their dissolved silica.

As already stated, neutral and alklaine waters are characteristic of red rocks, in which they promote the soda process and dissolve silica. In areas of the rapid decomposition of organic matter, the environment turns more acid because of the carbon dioxide generated. Acid barriers develop in such places and produce, for example, the silication of wood.

Let us also recall that limestones become silicated when the pH is lowered at the interphase of water and oil due to the oxidation of petroleum compounds to carbon dioxide.

The evaporate barrier develops in areas where the evaporation of ground waters is intense and where various oxysalts are deposited. Such barriers are represented by gypsum strata, caliche soils, saline efflorescences, etc. Concentration by evaporation was discussed in Chapter 11.

Sorptive barriers arise where the ground waters come in contact with rocks containing sorbents. The negatively charged sorbents —clays, peats, or coal—adsorb metallic cations. Such barriers are particularly common in the vicinity of oxidizing sulfidic ore deposits.

Adsorbents, like bauxite or brown ironstone, are positively charged, and therefore attract anions. This mechanism explains such impurities in brown ironstones as arsenic, vanadium, or phosphorus. Sorptive barriers are instrumental in the formation of the secondary dispersion haloes (Chapter 5).

A special variety of physicochemical barrier develops as a result of drastic changes in pressure and temperature. This type of barrier conditions the gas regimen of ground waters. Carbonated waters, ascending along faults, deposit calcite in the form of concretions or vein fillings:

$$\frac{Ca\,(HCO_3)_2}{\text{dil. sol.}} \rightarrow \frac{CaCO_3}{\text{solid}} + H_2O + \frac{CO_2}{\text{gas}} .$$

In karst regions, a thermodynamic barrier develops in places where ground waters come out on the day surface. Such barriers are marked by travertine deposits.

An illustrative example was furnished by Kaplanyan (1962), who investigated such phenomena in the Agartsin canyon in the Gugarts

Mountains of northern Armenia. A spring, which eventually becomes the left tributary of the Axtafa River, flows from under a limestone cliff at the head of that canyon. Travertine tuff has deposited around this spring and also occurs in several other places along the stream. These tuffs mark former thermodynamic barriers. Contrary to the opinion of some geologists, these tuffs were formed by supergene processes and are not the products of thermal springs.

Such deposits of calcium carbonate are also characteristic of soils in which they become illuvial carbonate layers.

A rise in temperature brings about a decrease in the solubility of carbon dioxide and the precipitation of calcium carbonate from waters carrying calcium bicarbonate. Calcareous layers are formed at such day surface. Eventually, the water-bearing strata become clogged with calcite (Kovda, 1946). As pressures rise and temperatures drop, other physicochemical barriers may develop.

Biologic barriers are a result of biogenic accumulation (Chapter 1). The vegetation covering the dry land is an example. This covering primarily accumulates C, O, H, N, and, to a lesser extent, aqueous migrants.

The humic layer of the soils, which locally concentrates some ore elements, is another example. Colonies of sulfur bacteria and other microorganisms are special types of biologic barriers. Their peculiarities should be reckoned with during geochemical prospecting.

A number of economic mineral deposits are formed at some biologic barriers, such as deposits of peat, coal, or perhaps some metallic minerals. Another function performed by biologic barriers is the sorting out of chemical elements. Soil fertility and the evolution of organisms are also conditioned by biologic barriers.

There are also geologic types of geochemical barriers. The same geochemical barrier may develop under different geological conditions; for example, the same barrier may develop in a marsh soil, in an oil field, or in an aquifer. Different as these processes might be in details, they will have something in common — the reducing, H_2S environment. Since this is true of all the barriers, it is convenient to speak in terms of "geologic types of geochemical barriers." A detailed study of these types has a definite practical value, since it leads to an understanding of natural phenomena and affords interpretations by analogy. An investigation of the geochemical barriers in soils casts light on the phenomena taking place in the water-bearing strata. A study of catagenetic ore deposits helps in the study of soil formation and soil weathering. A comparative study of the different modifications of the barrier in different environments is a powerful tool of the scientific investigation.

So far, the various types of the geochemical barriers have been considered separately, in order to stress their individual characteristics. In actual supergene environments, several geochemical barriers may be superimposed. For example, a thermodynamical barrier coincides in space with the oxygen barrier in places where the carbonated hypogene waters come out on the day surface. In such places, ferric hydroxides precipitate out of the gley phase of those waters in addition to the $CaCO_3$ which precipitates at the oxygen barrier. Biologic and sorptive barriers may coexist in the same humic layer of a soil.

Dobrovol'skii (1957) wrote: "A flood plain may be regarded as a barrier which prevents the removal of material from the banks and into the stream and which accumulates these materials in the swamps above the terraces." The accumulation of compounds of Fe, Ca, Mn, P, N, etc., on flood plains is conditioned by drastic changes in the contents of CO_2 and O_2, and, therefore, in pH and Eh.

Actually, one can speak of such complex barriers as biosorptive, oxygen-thermodynamic, etc.

EPIGENETIC CONCENTRATION OF ELEMENTS DURING CEMENTATION (APPROXIMATE CALCULATIONS)

Epigenesis often results in the cementation of water-permeable strata by the substances dissolved in waters. Calcareous, gypsiferous, and cupriferous sandstones are examples. Such processes are important to the formation of various ores including those of Cu, V, and S.

Let us now consider the quantitative implications of the epigenetic concentration of elements during the cementation of an aquifer. Let us assume that a reducing geochemical barrier exists in that aquifer (Fig. 57), and causes the precipitation of some dissolved substances. Gradually, a lens of a cemented rock develops. Let us say that this cement contains an ore element X, which is concentrated in the waters before the barrier at a_1 g/liter. Some

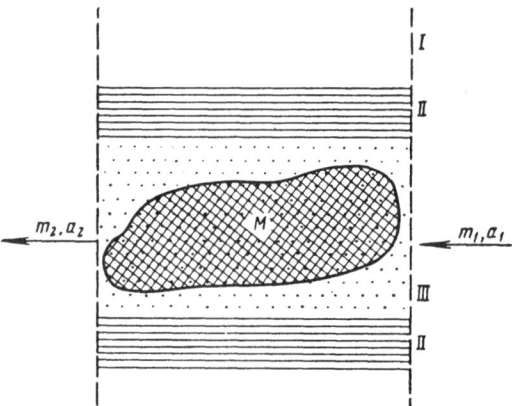

Fig. 57. The epigenetic concentration of elements in the vicinity of a geochemical barrier. (I) A geochemical barrier; (II) impermeable rocks; (III) the water-bearing aquifer.

of this compound will be stopped by the barrier and the remainder, a_2, will continue migrating below the barrier. Then, the amount $(a_1 - a_2)$ will precipitate at the barrier out of every liter of the circulating water.* Correspondingly, the content of element X above the barrier is m_1 g/liter and below the barrier it is m_2 g/liter. Let us further assume that the m_2 is very small in relation to m_1, i.e., that most of the X content is left at the barrier. Let M be the mass of the original rock, and   be the number of liters of water which pass through the barrier in the time t. Then, the total quantity of the element X which fell out of solution will be $(m_1 - m_2) Q$ g.

The mass of the cement formed by the precipitating components will be $(a_1 - a_2) Q$ g. The total mass of the cemented rock will be $M + (a_1 - a_2)Q$. The content, nX, of the element X in the cemented stratum will be:

$$n_X = \frac{(m_1 - m_2) Q \cdot 100}{M + (a_1 - a_2) Q}. \qquad (1)$$

Obviously, the value of $Q(a_1 - a_2)$ cannot be larger than the mass of the cement completely filling the rock pores. This limiting value is KM, where K is a coefficient depending on the rock porosity and the specific gravity of the cement.

*Actually, the concentration below the barrier may be larger than a_2 because of interstitial evaporation or some other upgrading factor. The important fact is, nevertheless, that $(a_1 - a_2)$ g /liter are being precipitated at the barrier.

Taking the rock porosity to be from 30% to 40% and the specific gravity of the cement to be slightly greater than that of the rock, the value of K will be between 20 and 50. In other words,

$$Q_{max}(a_1 - a_2) = KM \quad \text{or} \quad Q_{max} = \frac{KM}{a_1 - a_2}. \tag{2}$$

Substituting (2) into the equation (1) and making the necessary transformations, we have:

$$n_X = K \frac{m_1 - m_2}{a_1 - a_2}. \tag{3}$$

Thus, the epigenetic concentration of chemical elements in water-permeable rocks is directly proportional to the difference in their concentrations on either side of the barrier and inversely proportional to the total mass of substances precipitated at the barrier. These calculations do not take into account the effects of metasomatic replacements.

In many cases, the a_2 will be so small as to be negligible. Then equation 3 can be modified:

$$n_X = K \frac{m_1}{a_1 - a_2}. \tag{4}$$

In other words, the higher is the element's content in waters reaching the barrier, the greater will be its concentration within the barrier. However, the concentrations of other elements should be kept at a minimum if the element X is to make ore. Most favorable for the formation of ore are feebly mineralized, unsaturated solutions which precipitate little else at the barrier in addition to the ore.

Let us determine the copper content imparted to a sandstone by an infiltrating water carrying 0.5 g/liter (a_1) of the total mineralization, which includes $5 \cdot 10^{-5}$ g/liter Cu (m_1). Let us assume that half of the dissolved substances are left at the barrier, i.e., that $a_2 = 0.25$ g/liter.

Substituting these data into equation 4, we obtain:

$$n_{Cu} = K \cdot \frac{m_{Cu}}{a_1 - a_2}; \quad n_{Cu} = 40 \cdot \frac{5 \cdot 10^{-5}}{0.5 - 0.25} = 0.008\%.$$

This process cannot produce any copper ore.

What copper content should the infiltrating water have to produce a 1% Cu ore? We calculate

$$m_{Cu} = \frac{n_{Cu}(a_1 - a_2)}{K} = \frac{1 \cdot 0.25}{40} = 6.2 \cdot 10^{-3} \text{ g/liter}.$$

Such a high copper content is improbable in natural waters. Let us diminish the value of $(a_1 - a_2)$ by assuming a_2 to be 0.45 g/liter; then, substituting the values of 0.5 and 0.45 for a_1 and a_2 in the given equation, we obtain $1.2 \cdot 10^{-3}$ g/liter. This figure is only slightly higher than the copper contents of ground waters just outside ore deposits.

Consequently, the waters must be enriched in copper and low in other contents in order to form an ore deposit of the cupriferous sandstone type. A 1% copper ore may be formed by waters containing $2.5 \cdot 10^{-4}$ g/liter Cu if only 0.01 g precipitates at the barrier from the 0.5 g/liter of total dissolved substances.*

*All these calculations serve only to illustrate the mechanisms, since they deal with idealized cases. In reality, much of the ore is emplaced in porous sandstones by means of metasomatic replacement. Caution should be exercised when making such calculations on actual ore deposits.

Chapter 14

The Development of Epigenetic Processes Through Geologic History

The variation with time in the factors of the migration of chemical elements has influenced the development of epigenetic processes. Thus, the continents of the Precambrian period were characterized by a much smaller mass of living matter than the modern continents show. Hence, organic matter was less important then than it is now.

Although the accumulation of living matter has been prominent through geologic history, its intensity has varied from period to period. For example, there was more living matter in the Upper Triassic and Lower Jurassic periods — the epochs during which climates were humid in many parts of the Soviet Union — than in the preceding or following periods. Therefore, that epoch was noted for gley formation in carbonate-free environments. It may be that in some geologic periods organisms were more numerous than in this present time of high mountains and glaciated regions.

The accumulation of dead organic matter has probably always gone on. As time passed, new masses of peat, petroleum, and gas, and disseminated carbon were buried. Although the disseminated carbon comprises only a small fraction of one percent of the rock mass, the total mass of the disseminated carbon is more than 1000 times greater than the mass of the combined reserves of mineral fuels. All these organic substances constitute the source of energy for microorganisms, which extract oxygen from ground waters and enrich these waters with carbon dioxide. Microorganisms are a factor in the development of the reducing environments and the corresponding epigenetic processes.

Therefore, the quantity of organic matter buried in the earth's crust was proportional to the development of photosynthesis. In turn, this led to the greater microbiologic destruction of those substances. The geochemical activity of ground waters increased as these were becoming richer in Co_2, H_2S, CH_4, etc.

The outcome of the activity of living organisms was the accumulation of oxygen in the atmosphere and the removal of carbon dioxide from it. During their collective existence, plants have practically depleted the atmosphere of its carbon dioxide. This should have influenced the course of epigenetic processes. Over billions of years, the relative contents of the chemical elements in the earth have been changing (Saukov, 1958). As a result of radioactive decay, the quantities of uranium and potassium have been diminishing, and the contents of lead, argon, strontium, and other daughter elements have been increasing. Such changes should also have affected the course of epigenetic processes. It is possible that some processes operated in the past which are now nonexistent. This may be particularly true of the Precambrian and Paleozoic periods. Much caution should be exercised when applying the principle of uniformitarianism to ancient epigenetic changes. The Eh values may have been quite different in the past. This applies especially to the Lower Archaeozoic era, when there was no life on earth. The atmosphere lacked oxygen at that time. The behavior of elements like Se, Cr, U, and V had to be different then. If the Eh ranges were lower, then the conditions were less favorable for the migration of those elements. Possibly, those elements did not form any oxysalts. Chromium and vanadium probably always remained trivalent and selenium was probably always in the form of insoluble selenides. Uranium was always tetravalent.

At the present time, these elements can form soluble compounds under the proper conditions, this fact accounting for the high mobility of their oxy-acid radicals (Perel'man, 1963).

On the other hand, if in past geologic epochs the conditions permitted higher Eh values than at present — on the order of 1 V, for example — then the mobility of these elements must have been greater. These elements could then have migrated in environments in which they are now immobilized.

While the development of individual epigenetic processes is progressive and irreversible, the course of the processes, taken in groups, is cyclical. The succession of arid and humid climates is an example.

During epochs characterized by mountain building, ground waters become more dynamic, their zones of active circulation thicken, and the limit of oxidation is lowered. In general, the role of reducing processes decreases and that of oxidizing processes increases.

The sulfatic process gains prominence in the vicinity of sulfide veins. The spread of arid climates makes the development of cal-cium-carbonatic and chlor-sulfatic processes possible. The role of organic matter is reduced in such places.

In regions of a weakened tectonic regimen, descending oxygen-ated waters aid the development of reducing barriers. This re-sults, for example, in the formation of uranium deposits.

Due to the intense volcanism which accompanies mountain building, large quantities of carbon dioxide, chlorine, sulfur, and other nonmetals are added to the atmosphere. An increase in content of carbon dioxide in the atmosphere favors an increase in vegetation, coal-making, and gley formation.

Thus, the epochs of the Caledonian, Variscian, and Alpian oro-genies were never identical, although they had many features in common. We can thus speak of the cyclical recurrence of epi-genetic processes without having to assume that these cycles are reversible.

On the other hand, epochs of quiescent tectonics were charac-terized by the increased role of living matter in epigenetic processes. Humid climates prevailed. Reducing processes pre-dominated over oxidizing ones, and the lower limits of oxidation rose.

All regions of the earth surface must have repeatedly under-gone such changes. As a result, the effects of different and seem-ingly incompatible processes have become superimposed. For example in Central Asia, Lower Paleozoic shales are impregnated with acid gley which must be of a Mesozoic age.

There are reasons to believe that superpositions of processes have been quite common and diversified.

Rukhin (1956) places much weight on tectonic movements as the causes of such superimpositions, and introduces the concepts of progressive and retrograde epigenesis. A progressive epi-genesis develops during subsidences, when rocks sank to regions of higher pressures and temperatures. Silica begins to replace carbonic acid in the course of silication and sericitization. In the Cretaceous rocks of the Fergana, the dolomitization and the dehy-

dration of gypsum to anhydrite is another manifestation of progressive epigenesis.

A retrograde epigenesis develops during tectonic uplifts, when
rocks are placed into environments of low pressures and temperatures. The calcitization, de-dolomitization, and hydration of anhydrite to gypsum are typical.

Chapter 15

Conclusions

THE UNIQUENESS OF MIGRATION

We find different stages of physicochemical, mechanical, or biogenic migration in the history of every chemical element. These modes of migration are closely interrelated, the role of any given process varying with the element.

Biogenic processes may play a very important role in the history of some elements, such as potassium or phosphorus. Others, like sodium or chlorine, are mainly transported in the physicochemical manner. Still others, like tin or platinum, are transported mechanically.

The role of an individual process also varies with the environment. For example, iron migrates mechanically in deserts, but is transported in solution in swamps of the tundra and the taiga. Most elements are sorbed by living organisms, and are then transported either in solution or mechanically.

Therefore, all attempts to consider but one mode of migration and ignore the others are erroneous.

GEOCHEMICAL TYPES OF EPIGENETIC PROCESSES IN NATURAL MEDIA OF THE SUPERGENE ZONE

As already shown, a relatively small number of epigenetic processes characterize the supergene zone. The same process may develop in different media — soils, eluvium, deluvium, aquifers, etc. In other words, the same epigenetic processes may develop under different geologic conditions. The question of prototypes is based on the above statement. Certain processes may be taken as prototypes of the migration or of the concentration of a given element. For example, the saline layer of soils may be considered as the prototype of soda catagenesis. It could be used as a model of some ore deposition.

The availability of soils and the fact that various processes are easily observable in soils at the present time make these soils valuable media to an experimenter.

THE GEOCHEMICAL CLASSIFICATION OF THE NATURAL MEDIA OCCURRING IN THE SUPERGENE ZONE

The principles discussed above for classifying epigenetic processes may be utilized for the geochemical classification of soils, the weathered crusts, aquifers, or continental sediments. It should be kept in mind, however, that a given process may not be typical of all the media in the supergene zone.

The author has advanced a geochemical classification scheme for weathered crusts which is based on the concept of the mobility of the components and on the theory of typomorphic elements (Perel'man, 1957, 1962). Various types of crusts (= eluvium) are recognized on the basis of the most mobile crust components and of the chemical composition of the percolating waters. The carbonatic and the acid gley types of crusts, which develop in reducing environments, are being designated as "types" for the first time. The principal geochemical types of weathered crusts are as follows (based on the same principles as Table 27; the typomorphic elements, ions, or compounds are given in parentheses):

The oxidizing series

1. Sulfatic crust (SO_4^{2-}, H^+)
   a. Oxidation zone of sulfidic ore deposits
   b. Eluvium of pyritic clays and shales
2. Acid crust (H^+)
   a. Moderately leached
   b. Strongly leached
   c. Quartzose
3. Carbonatic crust (Ca^{2+})
   a. With gypsum predominant
   b. With calcium carbonate predominant
   c. Not containing $CaCO_3$, but containing Ca and Mg in the sorbent complex

4. Saline crust (Na^+, Cl^-, SO_4^{2-})

The reducing series

5. Acid gley crust (H^+, Fe^{2+})
6. Carbonatic gley crust (Ca^{2+}, Fe^{2+})

Such an approach is suitable for the classification of the continental sediments as well. Geochemically, some continental sediments resemble the eluvium very closely, while others differ considerably. Thus, in many cases, the deluvium and proluvium are of the same type as the eluvium, i.e., they are acid, carbonatic, etc. On the other hand, lacustrine and fluvial alluvial sediments are quite different from the corresponding crust.

Therefore, it is only logical to differentiate between subaerial and subaqueous sediments. The first closely resemble the crust, but the second do not. The classification scheme for the basic geochemical types of continental sediments is:

I. The oxidizing series

1. Acid sediments (H^+)
2. Carbonatic sediments (Ca^{2+})
3. Chlor-sulfatic sediments (Na^+, Cl^-, SO_4^{2-})

II. The reducing gley series

4. Carbonate-free gley sediments (H^+, Fe^{2+})
5. Carbonatic gley sediments (Ca^{2+}, Fe^{2+})
6. Chlor-sulfatic sediments (Na^+, Cl^-, SO_4^{2-})

III. The reducing H_2S series

7. Saline-sulfatic sediments (Na^+, Cl^-, SO_4^{2-}, H_2S)

The geochemical classification of soils and aquifers still needs to be worked out.

CRITERIA FOR THE RECOGNITION OF EPIGENETIC PROCESSES

Both positive criteria, which identify an epigenetic process, and negative criteria, which establish its absence, are given in Table 32.

Table 32. Identification of Epigenetic Processes

	Oxidizing					
	Sulfatic	Acid	Neutral carbonatic	Chlor-sulfatic	Gypsiferous	Soda
Alteration of rocks and soils. Epigenetic minerals	Formation of "iron caps"; bleaching. Accumulation of sulfates of Fe^{3+} and Al^{3+} (alums, jarosite, etc.) and of heavy metals. Local accumulations of arsenates, phosphates, carbonates, and silicates of the same metals	Formation of clay minerals, opal, chalcedony. Accumulation of hydroxides of Fe and Al. In soils, the formation of podzols and the absence of carbonates of Ca and Mg	Solution cavities. Secondary calcite in veins. Calcite cement in sandstones Red clays in limestones	Salt karst. Gypsum and oxy-salts in rocks. Gypsum, halite, mirabilite, and other soluble minerals. Dedolomitization.	Karst. Gypsum, celestite	Silication of rocks. Corrosion of grains of quartz and silicates. Synthesis of palygorskite and chrysocolla. Albitization. Migration and concentration of humates, U, V, Mo, Cu, etc. Pseudomorphism of calcite after gypsum
Criteria exclusive of the evolution of the given epigenetic process (no superpositions considered)	Absence of sulfides and native sulfur in unaltered rocks. Absence of oxysalts of heavy metals	Carbonatization	Absence of carbonates in rocks; intense migration of humates, iron, and aluminum			Gypsification of rocks; redistribution of iron

Table 32 (continued)

| | Gley-forming; reducing without hydrogen sulfide | | | | | Reducing, H$_2$S | |
	Carbonate-free gley	Gley-carbonatic	Oxysalt gley	Gypsum gley	Soda gley	Oxysalt-sulfidic	Soda-H$_2$S
Alteration of rocks and soils. Epigenetic minerals.	The gley impregnation of rocks and soils: bluish gray, white, ocherous colors. Signs of migration of iron. Secondary minerals: vivianite, siderite, ferrosilicates. The iron impregnation of sandstones and conglomerates. Alternation of bluish-gray and red members. Vari-colored; contacts of clays and alevrolites with sandstones and conglomerates	The gley impregnation of rocks and soils. Blue-gray, white, or ocherous colors. Alternation of blue-gray sands and gravels with red alevrolite and clays. Carbonatization of rocks. Black manganese hydroxides	Impregnation with gley; association with gypsum and soluble salts	Impregnation with gley in association with gypsum	Impregnation with gley; signs of soda process	Accumulation of gypsum and soluble salts; formation hydrotroilite, other epigenetic sulfides; H$_2$S; black color of soils and spots in rocks	Bleaching by oxidation of organics; pyrite; H$_2$S; silication
Criteria inclusive of evolution of the given epigenetic process (no superposition considered)	Uniform red or brown color of rocks; no carbonatization	Uniform red or brown color of rocks; carbonatization	Pyrite, hydrotroilite in considerable amounts		Pyrite, hydrotroilite in considerable amounts. Gypsum	Uniform red or brown color of rocks. No sulfates. Traces of gley	Uniform red or brown color of rocks. No sulfates. Traces of gley, gypsum

GEOCHEMICAL TYPES OF REDUCING
ENVIRONMENTS

The concept of two principal types of environments, the oxidizing and the reducing, has become firmly rooted in geochemistry.*

It has been customary to measure the reducibility of an environment in terms of the Eh.† Nevertheless, the geochemical limits of reducing environments do not depend so much on the Eh values as on the presence of hydrogen sulfide and its anions (Chapter 11).

The Eh and pH values remaining the same, the geochemical conditions vary with H_2S content. This is particularly evident in the case of iron. Iron may be quite mobile in strongly reducing environments or it may be practically immobilized, all depending on the absence or presence of hydrogen sulfide. The same holds true for most chalcophile elements which form insoluble sulfide ores.

These considerations as well as the material presented in the preceding chapters show that the most important factor in the development of reducing environments is H_2S and not the Eh value. Two types of reducing environments should be distinguished: (1) the gley environment without H_2S, and (2) the H_2S environment. The first of these has been known under different names. It has been called the peaty-colloidal horizon in pedology, the aiderite facies in sedimentology, and the ferruginous waters in hydrogeology. The commonness of gley in sedimentary rocks has been established in our investigations.

Until recently, it had never occurred to geologists to think of the gley itself, all their attention being occupied with phenomena of the secondary importance. We consider the gley environment to be one of the principal geochemical environments, and want to stress its role in the migration of elements.

THE GEOCHEMICAL CLASSIFICATION OF NATURAL WATERS

Two different tendencies are noted in existing classification schemes for natural waters. Some authors (such as O. A. Alekin,

---

*The redox conditions of the supergene zone have always commanded much attention in geochemistry, commercial geology, sedimentology, and hydrogeology. See the earlier works of Pustovalov, Shcherbina, and Fersman or more recent works of Shcherbakov, Germanov, Garrels, Krauskopf, etc.

†On the basis of the Eh, Popov and others (1963) recognized four environments: oxidizing, weakly reducing, reducing, and strongly reducing. A similar approach was used by Strakhov, Teodorovich, Shcherbakov, and others.

S. A. Shchukarev, and V. A. Sulin) base their classifications on the ionic content of waters, while others (V. I. Vernadskii or A. M. Ovchinnikov) classify waters on the basis of their gas content.

Ovchinnikov divided natural waters into three groups: (1) oxidizing, with oxygen, nitrogen, carbon dioxide, etc.; (2) reducing, with hydrogen sulfide, methane, and carbon dioxide, but without oxygen; and (3) metamorphic, with carbon dioxide, nitrogen, etc.

Waters of the supergene zone belong to the first two groups. In keeping with the subdivision of the reducing environments, Ovchinnikov's second group may be divided into (a) gley-forming and (b) H_2S-bearing. Since it has already been shown (Chapter 9) that the taxonomic significance of aeolean migrants (gases) is greater than the significance of aqueous migrants (ions), we prefer to use the gas content as the basis for classification.

We propose the division of natural waters into three groups:

1. Oxidizing waters, containing either free oxygen or some strong oxidizer, such as ferric sulfate. All stream waters, most marine and oceanic waters, all soil waters, and most other ground waters from zones of active circulation belong here.
2. Gley waters, which contain little or no oxygen or hydrogen sulfide and are capable of reducing ferric iron to the ferrous state. Carbon dioxide, methane, and other hydrocarbons are prominent among gases. This type includes many soil and ground waters from zones of active and obstructed circulation as well as swamp waters of humid climates.
3. Hydrogen sulfide waters containing S^{2-} and HS^- ions. These waters are not favorable to the migration of iron and chalcophile elements. Although these waters are typically from zones of obstructed circulation, some waters may come from blocks with active circulation.

We conclude that the gley medium is a more important factor for purposes of the classification than either the type or the amount of mineralization. All carbonatic, chloridic, and sulfatic waters may be classed together in one group as long as they contain no hydrogen sulfide. This "gley type" may be further subdivided into (a) fresh gley water, (b) saline gley water, and (c) brine. The oxidizing and the H_2S waters may similarly be subdivided into classes and each class may be subdivided further according to the nature of the constituent ions.

Dissolved organic matter of the humus type (fulvins, etc.) is an important constituent of many natural waters. Its content may be considerable. There are many stream and ground waters of this type in humid tropics and the tundra. The Rio Negro of the Amazon drainage system is an example. The geochemical significance of dissolved organic matter is extremely great, although it has always been overlooked in taxonomy.

Waters having the same ionic constituents and equal in total ion content but differing in their content of organics are distinctly different geochemical entities. In many cases, the taxonomic significance of the dissolved organics is greater than the significance of the dissolved ions. For this reason, a geochemical classification of natural waters should take into account the contents of the dissolve organic substances. The design of such a classification is one of the problems of geochemistry.

VERTICAL HYDROCHEMICAL ZONING IN GROUND WATERS

Like any other theoretical generalization, the theory of the vertical zoning of ground waters exerts some influence on all the allied fields of science. The theory, which was worked out by N. K. Ignatovich and other hydrogeologists, has for example, influenced Rukhin's work in sedimentology.

The theory of zoning bears directly on geochemistry. Such problems as the formation of exogene epigenetic ore deposits and the formation of the low-temperature hydrothermal ore deposits are probably conditioned by the vertical hydrochemical zoning. Therefore, the geochemical principles for the identification of such zones are of the utmost importance.

The amount of total mineralization and the corresponding ionic composition have been used by Germanov and others (Chapter 8). Nevertheless, the role of oxidation−reduction environments has never been utilized to the fullest extent.

As already shown, such conditions determine the geochemical type of the water. We share in the opinion that there should exist various types of water zoning which depend on the geologic structure of the region. At least two types of vertical hydrochemical zoning should be recognized.

The first type of vertical hydrochemical zoning is characterized by the gley waters. It closely resembles Germanov's "strata without any reacting organic matter." The difference is due to the conviction that the first type should be characterized by the presence or absence of organic matter dissolved in water, and not just contained in host rocks.

If the sulfate content of the medium is low enough, the organic matter of the host rocks will not generate any hydrogen sulfide. Such is the case of the swamp waters of the north, which evolve no hydrogen sulfide, despite the organic matter accumulated in the enclosing rocks.

The uppermost zone of this Type I may be called a "zone of oxygenated waters." Iron does not migrate in this zone except in the oxidation of sulfides. The waters are slightly mineralized. According to Ignatovich, their circulation is active. The enclosing rocks are colored yellow, brown, or red.

Below this zone lies "a zone of cold gley waters"; these waters carry little or no oxygen, and are weakly mineralized. Their contents of bivalent iron and manganese increase as the gley forms. The water circulation is more or less obstructed.

Downwards, this zone grades into a "zone of thermal gley waters," which carry carbon dioxide, ferrous iron, etc.

This is a generalized scheme. In some cases, as in the swamps of the north, the uppermost zone of oxygenated waters may be entirely absent. Then, the cold gley waters begin at the very surface.

Each of these zones may be subdivided. Thus, a subzone of mobile manganese may be distinguished in the lower portion of the first zone. The second zone can be subdivided into "weak" and "strong" subzones. The third one may be differentiated into several chemical or thermal subzones.

While Ignatovich had designed his scheme for artesian basins, * ours is fitted to the regions of folded mountains. Since the igneous and metamorphic rocks of such regions are intensely fissured, the waters become slightly mineralized (i.e., they contain neither oxysalts nor organic matter).

The second type of vertical hydrochemical zoning includes a zone of hydrogen sulfide. Its uppermost zone contains oxygenated and slightly mineralized waters which circulate freely. The gley

---

*It was shown by F. A. Makarenko that the zones of active water circulation are potent in the folded regions. He subdivided those zones into subzones of local and regional drainage. Kaplanyan (1962) applied this to geochemical prospecting and showed that only the subzone of local drainage can be utilized in prospecting.

Table 33. Paleohydrochemical Reconstructions on the Basis of Epigenetic Rock Alteration

	Epigenetic Processes								
	Sulfatic	Acid	Neutral-carbonatic	Chlor-sulfatic	Soda	Carbonate-free; gley	Gley, carbonatic	Oxysalt-sulfidic	Soda-H_2S
Chemical character of water	pH under 3. Free H_2SO_4 and a small Cl^- content. Locally high content of Cu, Zn, Al, and other metals	pH 4 to 7. Slightly mineralized (under 0.5 g/liter) with Ca bicarbonate or silica	Neutral to alkaline, hard, with bicarbonates	Strongly mineralized and also brines	Alkaline	Slightly mineralized. Weakly acid. Bicarbonates of Ca, Fe, Mn. No oxygen	Slightly mineralized, hard, with bicarbonates of Ca, Fe, Mn. No oxygen	Mineralized. With H_2S and no oxygen	Alkaline. With H_2S
Examples	Paleogene waters of oxidation zones of the Urals. Triassic and Jurassic waters of pyrite shales of Central Asia	Ground waters of U. Triassic and Leyass of Urals, Kazakhstan, and Central Asia	Waters of limestones of various ages	Waters of rock salt deposits and their suites	Waters of red sandstones in different regions of the USSR (?)	Albian and Senonian in red rocks of Central Asia, or under bogs of Upper Triassic in Central Asia and Kazakhstan	Waters of the ancient aquifers in red rocks of different ages in Urals area, Kazakhstan, Donbass, and Central Asia	Ground waters of certain alluvial plains of Neogene in Central Asia	Waters of certain oil deposits which are now being destroyed

zone below is poorly developed if at all present. Still lower lies a zone of hydrogen sulfide and its anions. The environment is generally reducing. Geochemical barriers develop at the upper boundary of this zone and cause the deposition of uranium and various sulfides (Chapter 12). The waters are more strongly mineralized and their circulation is obstructed.

The H_2S content diminishes downwards as the water becomes more mineralized, loses its sulfatic content, and becomes stagnant. This zone can be called a zone of strongly mineralized gley waters.

This second type of vertical hydrochemical zoning develops in rocks which are low in organic matter. Hydrogen sulfide and its anions are introduced into the strata from outside sources.

These two types of vertical zoning do not cover all possible cases, and much remains to be ascertained. Needless to say, while the first type develops mostly in regions of mountain folding and the second on the platforms, each may occasionally develop in the territory of the other.

Studies of the epigenetic alteration of rocks permit the reconstruction of the chemical nature of ground waters of past geologic epochs. Flows of ground water, as we noted, left behind trails of epigenetically cemented rock. The chemical composition of these ancient waters may be inferred from the composition of the cement. The analysis of manganese hydroxides is particularly valuable in this respect, since these adsorb various metals. The trace elements found in black films of manganese hydroxides therefore indicate the composition of the ancient waters which circulated through the stratum. Various examples of such interpretations are given in Table 33. Unfortunately, this method does not date the events, and, therefore, it should be used in conjunction with some other method in order to establish the chronology.

SUPERGENE AND HYDROTHERMAL PROCESSES

Investigations of hydrothermal and supergene processes are only loosely interrelated. This is undoubtedly due to the different origin of the waters in these cases — juvenile in one, vadose in the other.

Aside from such inherent differences, these two groups of processes have certain features in common. They are controlled by the same laws governing the behavior of the elements in aqueous solutions. Certain relationships which were worked out for one group may be applied to the other group by analogy.

Korzhinsky, for example, used such analogies freely in his work on metasomatism, hydrothermal solutions, etc.

In making such analogies, we do not equate the geneses of the two types of mineral deposits, but only attempt to clarify the mechanisms of migration.*

The hypogene origin of hydrothermal solutions explains their predominantly reducing character. Within the supergene zone, such media may be: (1) gley media without H_2S, or (2) reducing media with H_2S. This statement is being repeated in order to stress the importance of hydrogen sulfide.

Thermal gley waters are common in the earth's crust and particularly in the regions of folding, or in the regions of rocks which are low in organic matter. Generally, such waters carry chalcophile elements in solution.

The composition of thermal gley waters varies with the temperature of the enclosing rocks. Carbon dioxide is their major constituent.

According to Ananyan (1963), Dzhermuk thermal waters in Armenia contain various metals in solution. Tkachuk (1960) has similarly reported on the thermal waters of southeastern Siberia. Undoubtedly, various types of thermal gley waters exist, such as neutral, weakly alkaline, strongly alkaline, etc.

The concept of gley should be a part of the theory of hydrothermal solutions.

Thermal H_2S waters are also common in the earth's crust and are also encountered occasionally in regions of folding. However, such waters are typical of rocks enriched in organic matter. Hydrogen sulfide is generated during the desulfurization reactions. Not infrequently, the content of waters is related to the frontal flexures of geosynclines, where they originate in ruptured oil and gas anticlines. Although these waters vary quite widely in composition, their content of chalcophile elements is generally very low.

*Lately, the origin of hydrothermal solutions and hydrothermal ores has received a number of different interpretations. While the earlier tendency was to identify hydrothermal activity with postmagmatic activity, more recent investigations indicate that thermal waters are common to all regions, even to those not affected by volcanic activity. Hydrogeologists have shown that in volcanic regions modern hydrotherms are dependent on atmospheric precipitation. Moreover, many believe that some hydrothermal ore deposits have never been related to regional volcanic activity. Thus, Ovchinnikov's investigations led him to believe that "some mineral deposits were formed in places where ancient waters dumped their chemical loads."

M. Konstantinov, A. A. Saukov, A. I. Germanov, and others have described various metalliferous deposits of nonmagmatic origin.

Without doubt, there should exist some transitional environments with waters which carry chalcophile elements in amounts
insufficient for the precipitation of sulfides. In such a case, only
a small part of metals would precipitate, in keeping with the principle of retardation. The remainder will continue to migrate.

The study of thermal gley waters should be approached with
the idea in mind that such waters are not ore-forming solutions.
Such a premise seems logical, since sulfide precipitates only in
places where gley and H_2S waters mingle. A proper geochemical
barrier develops in such places.

Speaking in a very general way, a certain regularity is noted in
the distribution of the two types of water in regions of folding.
Thermal gley waters develop in the central parts of geosynclines,
where the rocks are mostly igneous or metamorphic. Thermal
H_2S waters predominate on the periphery of such regions, where
the rocks are mostly sedimentary. In fact, H_2S is generated in
large amounts in deposits of oil and gas localized in ruptured anticlines on the peripheral front of a geosyncline.

Tectonic uplifts displace these zones. Gley waters begin to
migrate out. Meeting a hydrogen sulfide barrier, they precipitate
metal sulfides.

The question arises of whether or not the belts of sulfide ore
deposits could form in such a way, along a geosyncline which
stretches for hundreds and even thousands of kilometers. As
examples of this we cite the chalcopyrite belt of the Urals and the
Great Silver Belt of the Cordillerae, the latter stretching over
3000 km.

THE FORMATION OF MINERAL DEPOSITS SURROUNDED
BY SECONDARY DISPERSION HALOES

As shown in Chapters 10-13, many mineral deposits are
formed by epigenetic processes (Table 34). The same epigenetic
processes also destroy mineral deposits. Secondary dispersion
haloes are formed about the ore deposits as these are being
destroyed.

Geochemical barriers produce these haloes in unconsolidated
sediments and soils. The alkaline and sorptive barriers are particularly important in this respect.

It is well known that the main value of the geochemical method
of exploration lies in the detection of hidden ore deposits. Un-

Table 34. Mineral Deposits Formed by Epigenetic Processes

	Types of commercial mineral deposits	Examples
Epigenetic processes		
Sulfatic	Rich, oxidized ores of Cu, Zn, Pb, and other metals. Alum cappings of rock salt. Gypsum	Blyava and Mednorudyansk in the Urals, Shor-suh in Central Asia; gypsum deposits in SW Iran
Acid	Iron ores; siliceous ores of Ni; eluvial bauxites; kaolins; ores of rare elements, e.g., rare earths	Iron ore deposits of Cuba; nickel deposits in the Urals; bauxites of Arkansas (USA); kaolins of the Urals and the Ukraine
Neutral-carbonatic	Phosphorites deposited in solution cavities	Certain phosphorites of Belgium and the USA
Chlor-sulfatic	Salt brines	Khodzha-Mumyn and other brines of Tadzhikistan
Carbonate-free; gley	Small deposits of vivianite (blue phosphatic ores)	Swamps of Belorussia
Sulfato-sulfidic	Rich Cu ores in cementation zones of sulfidic ore deposits	Chalcopyrite ore deposits of the Urals, the Kounrad, and the western USA
Geochemical barriers		
Oxygen	Iron ores. Deposits of sulfur	Lake and bog iron ores of Karelia and Gaurdak; sulfur deposits of the Carpathian region
Reducing, H_2S	Uraniferous deposits; cupriferous sandstones (Cu, Ag, V, U, Zn, etc.)	Ambrosia Lake (USA); cupriferous sandstones of the western USA
Alkaline	Iron ores and siliceous nickel ores in solution cavities	Adapaev ore deposits of the Urals; Ufalei type of nickel ore deposits of Central Urals
Evaporate	Deposits of rocksalt, mirabilite, and other water-soluble salts	Brines of the Kara-bugaz

fortunately, the application of this method is limited to known aquifers and surface waters. In its present form, the method fails in dry rocks.

Not all rocks have once contained water. Where waters did percolate, they left behind trails of epigenetic alteration. A careful study might detect haloes in these alterations and lead to the discovery of ore bodies.

In this respect, various extracts, such as aqueous, saline, and alkaline fractions, of rocks and soils are useful.

METHODS OF STUDYING EPIGENETIC PROCESSES

A field investigation of epigenetic alteration involves mapping. Here we will discuss only methods applicable to the study of catagenesis, since other methods are well known.

Essentially, the methods of study of catagenesis consist of logging all traces of epigenetic alteration. This should also include unaltered members for comparison purposes.

Within the limits of the same aquifer, there exist definite paragenetic assemblages of epigenetic layers. Catagenetic profiles should be made in analogy with soil profiles. All changes in the mobile constituents — iron, manganese, and silica — should be carefully noted.

A detailed mineralogical analysis is necessary, particularly of the colloidal substances, which may have adsorbed some trace metals. This is preferably done spectrographically.

Another approach is through the experimental reconstruction of the presumed conditions by means of laboratory models. Very little has been published in this field (Priklonsky and Oknina, 1960; Evseeva, 1962).

The distribution geochemical types of catagenesis is then plotted on maps. Such maps are quite different from the usual ones. In a given region, rocks of different ages were often affected by the same epigenetic process. On the other hand, rocks of the same age were seldom affected by the same process in different regions. Therefore, such maps do not resemble other geologic maps. They show geochemical barriers, limits of oxidation, and various alteration zones, but not the types of original rocks. Generalized maps may be on a scale of 1:5,000,000 to 1:2,000,000, and more detailed ones on a scale of 1:100,000. Such maps should have some practical value in delineating the areas worth prospecting.

Publisher's Note

The following journals cited in this work are available in cover-to-cover translation:

Russian Title	English Title	Publisher
Doklady Akademii Nauk SSSR	Doklady Earth Sciences Section	American Geological Institute
Izvestiya Akademii Nauk SSSR: Seriya Geologicheskaya	Bulletin of the Academy of Sciences of the USSR: Geologic Series	American Geological Institute
Kolloidnyi Zhurnal	Colloid Journal	Consultants Bureau
Pochvovedenie	Soviet Soil Science	Soil Science Society of America

Bibliography

Alekin, O. A., Principles of Hydrochemistry. Gidrometeorizdat (1953).

Ananyan, A. L., "Hydrogeological and Geothermal Characteristics of the Dzhermuk Resort," A dissertation. Erevan (1963).

Andreev, Yu. K., and V. N. Godovikov, "On the Formation of Alkali Hornblendes in Permian Marls in the Dzhezkazgan Basin," Tr. GIGEM No. 31, Asbestos, a Mineable Mineral. Izd. Akad. Nauk SSSR (1959).

Antipov—Karataev, I. N., "Problems and Geographical Distribution of Salt Flats in the USSR," in: Upgrading of Salt Flats in the USSR. Izd. Akad. Nauk SSSR (1953).

Antipov—Karataev, I. N., and G. M. Kader, "On the Nature of Sorption by Clays and Soils," Kolloidn. Zh. Nos. 2 and 3 (1947).

Antipov—Karataev, I. N., and G. O. Tsyuryupa, "On Modes and Conditions of the Migration of Substances Through a Soil Profile," Pochvovedenie No. 8 (1961).

Babinets, A. E., Ground Waters of the Southwestern Russian Platform. Izd. Akad. Nauk UkrSSR, Kiev (1961).

Baghinskas, B., "Some Data on the Content of Mobile Microelements in Soils of the Lithuanian SSR," in: Microelements in Agriculture and Medicine, Fourth All-Union Conference, Ukr. Acad. Ag. Sci. Izd. UASKhN, Kiev (1962).

Bardoshi, D., and M. Bod, "A New Method of Measuring the Redox Properties of Sedimentary Rocks," Geokhimiya No. 3 (1960).

Bassett, H., "Silication of Rocks by Surface Waters," Am. J. Sci. Vol. 252, No. 12 (1954).

Batulin, S. G., "Evaporate Concentration of Rare Elements in Deserts and Steppes," in: Geochemistry of Steppes and Deserts. Geografgiz (1962).

Batulin, S. G., "Migration of Iron during Gley Impregnation in Lower Cretaceous Rocks in Southeastern Fergana," in: Problems of Geochemistry, Tr. GIGEM No. 99. Izd. Akad. Nauk SSSR (1963).

Becking, B., I. R. Kaplan, and L. Moore, "Variations of pH and Redox Potentials of Natural Media," in: Geochemistry of the Lithosphere. Russian translation, IL (1963).

Belyakova, E. E., A. A. Reznikov, L. E. Kramarenko, A. A. Nechaeva, and T. F. Kronidova, A Hydrochemical Method of Exploring Ore Deposits. Gosgeoltekhizdat (1962).

Beneslavskii, S. I., "Secondary Processes—The Most Important Factor in the Formation of Bauxites," in: The Weathered Crust, Part 4. Izd. Akad. Nauk SSSR (1962).

Beneslavskii, S. I., "Hydrogeological Regimen—The Most Important Factor in Bauxite Formation," in: The Weathered Crust, Part 5. Izd. Akad. Nauk SSSR (1963).

Birshtekher, E., Petroleum Microbiology. Gostoptekhizdat (1957).

Bogomyakov, G. P., and V. A. Nudner, "Thermal Waters of the West Siberian Artesian Basin and Their Practical Significance," Razvedka i Okhrana Nedr No. 9 (1963).

Borovskii, V. M., "On the Salt Exchange between the Sea and the Land and the Perennial Dynamics of Soil Processes," Pochvovedenie No. 3 (1961).

Brodskii, A. A., "Principles of the Geochemical Method of Exploration for Sulfide Ore Deposits," in: Mineral Resources. Nedra (1964).

Brusilovskii, S. A., "On the Migrational Modes of Elements and Natural Waters," Gidrokhim. Materialy Vol. 35 (1963).

Bughelskii, Yu. Yu., "Migration of Ore Components in Ground Waters in Mineralized Regions Located in Different Climates," Interdepartmental Conference on Hydrogeochemical Methods of Prospecting, Tomsk Univ. (1962).

Bughelskii, Yu. Yu., "On the Thermodynamics of Supergene Migration of Ore Components," in: Conf. on Hydrogeology and Eng. Geology, Sect. of Geochemistry. Moscow (1963).

Bushinskii, G. I., "On Diagenesis in the Genesis of Fire Clays, Sedimentary Iron Ores, and Bauxites," Izv. Akad. Nauk SSSR Ser. Geol. No. 11 (1956).

Chang Sheng, "The Contents and Migration of Boron, Iodine, Vanadium, Chromium, Manganese, Cobalt, Nickel, Copper, and Zinc in Certain Soils, Plants, and Waters of the USSR and Chinese Peoples Republic," A dissertation. Moscow (1962).

Chukhrov, F. V., The Oxidation Zone of Sulfidic Ore Deposits of Central Kazakhstan. Izd. Akad. Nauk SSSR (1950).

Chukhrov, F. V., Colloids in the Earth's Crust. Izd. Akad. Nauk SSSR (1955).

Degenhardt, Kh., "On the Geochemical Occurrence of Zirconium in the Lithosphere," in: Geochemistry of Rare Elements. Russian translation, IL (1959).

Dobritskaya, Yu. I., E. G. Zhuravleva, L. P. Orlova, and M. G. Shirinskaya, "Content of Microelements in Some Soils of the European USSR," in: Microelements in Agriculture and Medicine, Fourth All-Union Conference, Ukr. Akad. Ag. Sci. Izd. UASKhN, Kiev (1962).

Dobrovol'skii, V. V., "The Problems of the Theory of Soil Formation in Catchments of Streams of the Forest Zone," Vestn. Mosk. Univ. No. 1 (1957).

Dobrovol'skii, V. V., "Mineralogy and Geochemistry of New Formations in the Quaternary Strata of the Central-Russian Forested Steppe," A dissertation (1957).

Dobrovol'skii, V. V., "On Features of Supergenesis in Quaternary Deposits of Northern Kazakhstan," Geokhimiya No. 2 (1959).

Dobrovol'skii, V. V., "Typomorphic Zonal Manifestations of Supergenesis and Their Significance," International Geological Congress, XXI Session (1960).

Dobrovol'skii, V. V., "On the Geochemical Features of the Colloidal Fraction of Quaternary Sediments and Its Origin," Dokl. Akad. Nauk SSSR Vol. 147, No. 2 (1962).

Dobrovol'skii, V. V., "Supergenesis of the Quaternary Period and Its Geographical Aspects," A dissertation. Mosk. Gos. Univ. (1964).

Egorov, V. V., "Contamination of Soils with Soda in Southern Sin'tsian," Pochvovedenie No. 5 (1961).

Evseeva, L. S., and A. I. Perel'man, Geochemistry of Uranium in the Supergene Zone. Atomizdat (1962).

Fersman, A. E., Investigation of Magnesian Silicates. Izd. Akad. Nauk SSSR (1913).

Fersman, A. E., The Geochemistry of Russia. Izd. Akad. Nauk SSSR (1922).

Fersman, A. E., "The Geochemical Problems of Sulfur Hills in the Karakum Desert," in: The Sulfur Problem in Turkmenistan. Izd. Akad. Nauk SSSR (1926).

Fersman, A. E., Geochemistry. ONTI Vol. I (1933), Vol. II (1934), Vol. III (1937), Vol. IV (1939).

Friedland, V. M., "The Role of Weathering in the Formation of the Soil Profile and the Differentiation of the Soil Mass," Pochvovedenie No. 12 (1955).

Frolov, N. M., "The Thermal Zoning of the Hydrosphere in the Subsurface," Conference on Hydrogeology and Engineering Geology, Section on Mineral, Thermal, and Industrial Waters. Moscow-Erevan (1963).

Fundamental Chemical Features of Uranium. Izd. Akad. Nauk SSSR (1963).

Ganeev, I. G., "On the Feasibility of the Transfer of Substances in the Form of Complex Compounds," Geokhimiya No. 10 (1962).

Garrels, R., Mineral Equilibria at Low Temperature and Pressure. Harper, New York (1960).

Gedroits, K. K., The Theory of the Sorptive Capacity of Soils. Sel'khozgiz (1933).

Geissler, A. N., "On the Color of Varicolored Formations," Trans. All-Union Mineralog. Soc. Vol. 78, No. 2 (1949).

Genkin, A. D., "On the Corrosion of Quartz in Sulfidic Veins," Trans. All-Union Mineralog. Soc. No. 4 (1954).

Germanov, A. I., "Oxygen in Ground Water and Its Geochemical Significance," Izv. Akad. Nauk SSSR Ser. Geol. No. 6 (1955).

Germanov, A. I., "Basic Principles of the Hydrochemical Conditions of Formation of Infiltrated Uranium Deposits," Izv. Akad. Nauk SSSR Ser. Geol. No. 8 (1960).

Germanov, A. I., "The Role of Organic Matter in the Formation of Hydrothermal Sulfidic Ore Deposits," Izv. Vysshikh Uchebn. Zavedenii, Geol. i Razvedka No. 8 (1961).

Germanov, A. I., "Hydrodynamical and Hydrochemical Conditions of Formation of Certain Hydrothermal Ore Deposits," Izv. Akad. Nauk SSSR Ser. Geol. No. 7 (1962).

Germanov, A. I., S. G. Batulin, G. A. Volkov, A. K. Lisitsin, and V. S. Serebrennikov, "Some Relationships in the Distribution of Uranium in Ground Waters," in: Nuclear Fuel and Reactor Metals, Proc. II International Conference on Peaceful Utilization of Atomic Energy. Atomizdat (1959). English edition published by Pergamon, London-New York.

German–Rusakova, L. D., "Migration of Elements in the Oxidation Zone of the Blyavinsk Chalcopyrite Deposit in S. Urals," Tr. GIGEM No. 68. Izd. Akad. Nauk SSSR (1962).

Ginzburg, I. I., "Geochemistry and Geology of the Ancient Weathered Crust of the Urals. Part II, The Ancient Weathered Crust of Ultrabasic Rocks," Tr. Inst. Geol. Nauk, Akad. Nauk SSSR Vol. 81 (1947).

Ginzburg, I. I., "The Ancient Weathered Crust in the Territory of the USSR, Its Minerals and Properties," in: Transactions of the Session Dedicated to the Centennial of the Birth of V. V. Dokuchaev. Izd. Akad. Nauk SSSR (1949).

Ginzburg, I. I., A Contribution to the Theory of Geochemical Exploration for Nonferrous and Rare Metals. Gosgeoltekhizdat (1957).

Ginzburg, I. I., "Fundamental Problems of Formation of the Weathered Crust and Their Value in Exploration," Geol. Rudn. Mestorozhd. No. 5 (1963).

Ginzburg, I. I., and I. A. Rukavishnikova, "Pelicanites and Opals of the Kos-Shoku (N. Caucasus), a Comparison with Those of the Ukraine," in: The Weathered Crust, Part 1. Izd. Akad. Nauk SSSR (1952).

Ginzburg, I. I., and I. V. Vitovskaya, "Corrosion of Quartz in Clay of Hydromica-Montmorillonite Composition," in: The Weathered Crust, Part 2. Izd. Akad. Nauk SSSR (1956).

Ginzburg, I. I., A. A. Kats, and I. Z. Korin, "The Ancient Weathered Crust in Ultrabasic Rocks of the Urals. Part I, Types and Morphology of the Ancient Weathered Crust," Tr. Inst. Geol. Nauk, Akad. Nauk SSSR Vol. 80 (1946).

Glagoleva, M. A., "Modes of Migration of Elements in Stream Waters," in: Investigation of the Diagenesis of Sediments. Reports of the Committee on Sedimentary Rocks. Izd. Akad. Nauk SSSR (1959).

Glazovskaya, M. A., et al., Geochemistry of Landscapes and the Search for Ores in the Southern Urals. Mosk. Gos. Univ. (1961).

Goldschmidt, V. M., A Collection of Papers on the Geochemistry of Rare Elements. ONTI (1938).

Goleva, G. A., "Geochemistry of Ground Waters in Ore Deposits of the Western Ukraine," in: Problems of Geochemistry, Vol. I. Izd. L'vov. Inst. (1959).

Grigoriev, S. M., On the Formation and Properties of Mineral Fuels. Izd. Akad. Nauk SSSR (1954).

Grodzinskaya, K. P., "On Bacterial Transformations of Manganese in Soils," in: Microelements in Agriculture and Medicine, Fourth All-Union Conference, Ukr. Acad. Ag. Sci. Izd. UASKhN (1962).

Gurevich, M. S., "Rare and Disperse Elements in Artesian Waters," Byul. Vses. Nauchn.-Issled. Geol. Inst. No. 2 (1960).

Gurevich, M. S., "Role of Microorganisms in the Formation of the Chemical Composition of Ground Waters," Tr. Inst. Mikrobiol. Akad. Nauk SSSR, Vol. IX (1961).

Gurevich, V. I., and A. I. Pavlov, "On the Constant of Energy in Natural Waters," Conference on Applied and Engineering Geology, Geochemical Section, Moscow-Erevan (1963).

Gvozdetskii, N. A., The Karst. Geografgiz (1954).

Ivanov, M. V., "The Role of Microorganisms in the Formation and Destruction of Sulfur Deposits," Tr. Inst. Mikrobiol. Akad. Nauk SSSR Vol. IX (1961).

Kaleda, G. A., "Notes on the Problem of the Epigenesis of Sedimentary Rocks," Tr. Mosk. Geol.-Razved. Inst. Vol. 33 (1958).

Kaplanyan, P. M., "An Experiment in the Pedological—Hydrochemical Survey of Basalts," Izv. Akad. Nauk Armenian SSR Vol. XV, No. 4 (1962).

Karapet'yants, M. Kh., Chemical Thermodynamics. Goskhimizdat (1953).

Karapet'yants, M. Kh., and M. L. Karapet'yants, Tables of Certain Thermodynamic Properties of Various Substances. Goskhimizdat (1961).

Kartsev, A. A., Hydrogeology of Petroleum and Gas. Gostoptekhizdat (1963).

Karyakin, L. M., and I. N. Remezov, "Alunite Cements in Chasov—Yatsk Sands (Donbass)," in: Problems of Mineralogy of Sedimentary Formations, Chapters 3 and 4. Izd. L'vov. Gos. Inst. (1956).

Kassin, I. G., "Hydrodynamic Zoning in Deep-Seated Aquifers of the Artesian Basin Adjoining the Eastern Caucasus," Dokl. Akad. Nauk SSSR No. 5 (1962).

Kaurichev, I. S., E. V. Kulakov, and E. M. Nozdrukova, "On the Formation of Inorganic Compounds in Soils and Their Migration," Pochvovedenie No. 12 (1958).

Keller, V. D., "Principles of Chemical Weathering," in: Geochemistry of Lithogenesis. Russian translation, IL (1963).

Khod'kov, A. E., "Peculiarities of Formation of Weathered Crust of Halogenated Formations," Vestn. Leningr. Univ. Ser. Geol. i Geogr. Vol. 6, No. 1 (1963).

Kholodov, V. N., et al., "On the Epigenetic Zoning of the Uranium Mineralization in Oil-Bearing Carbonate Rocks," Izv. Akad. Nauk SSSR. Ser. Geol. (1961).

Konstantinov, M. M., Origin of Stratified Ore Deposits of Zinc and Lead. Izd. Akad. Nauk SSSR (1963).

Kopeliovich, A. V., "Epigenetic Processes in Arenaceous Rocks in the Dniestr Region," A dissertation. Mosk. Gos. Univ. (1962).

Korzhinsky, D. S., Physicochemical Basis of the Analysis of the Paragenesis of Minerals. Consultants Bureau, New York (1959).

Korzhinsky, D. S., Theory of Mineral Formation. Izd. Akad. Nauk SSSR (1962).

Kotyukov, I., Physical Chemistry, Vol. I. Tomsk. Gos. Univ. (1963).

Koval'chuk, A. N., L. P. Sokolova, A. Yu. Gil'miyarov, and E. M. Srodnykh, "Ferruginous Waters of the Eastern Slope of the Central Urals as Indicators Leading to Deposits of Iron, Fire Clay, and Bauxites of the Alapau Type," Conference on Hydrogeology and Engineering Geology, Geochemical Section. Moscow (1963).

Kovalevskii, A. L., "On Biogenic Accumulation of Chemical Elements in Soils," Izv. Sibirsk. Otd. Akad. Nauk SSSR No. 9 (1962).

Kovda, V. A., The Origin of Saline Soils and Their Regimen. Izd. Akad. Nauk SSSR (1946).

Kovda, V. A., I. V. Yakushevskaya, and A. N. Tyuryukanov, Microelements in Soils of the USSR, Mosk. Gos. Univ. (1959).

Kozlov, V. P., and G. A. Gladyshev, "An Experiment in Geochemical Correlation in the Bakhmut Basin," in: Geochemical Methods of Exploration for Oil and Gas. Izd. Akad. Nauk SSSR (1960).

Krainov, S. R., and M. Kh. Korol'kova, "Microelements in Mineral Waters of the Little Caucasus," Conference on Hydrogeology and Engineering Geology, Geochemical Section. Moscow (1963).

Kramarenko, L. E., and I. I. Prizrenova, "Denitrogenating Bacteria in Sulfidic Ore Deposits and Their Identification in the Field," Tr. Vses. Nauchn.-Issled. Geol. Inst. Gosgeoltekhizdat (1961).

Kramarenko, L. E., R. I. Teben'kova, and I. I. Prizrenova, "Sulfur Bacteria in Ground Waters in Ore Deposits of Rare Metals in Dzhungar Region of the Balkan Metallogenic Province and Their Significance," in: Materials of Regional Hydrogeology. Gosgeoltekhizdat (1961).

Krasnikov, V. I., Rational Search for Ore Deposits. Gosgeoltekhizdat (1959).

Krauskopf, K., "Sedimentary Ore Deposits of Rare Metals," in: Problems of Ore Deposition. Russian translation, IL (1958).

Krauskopf, K., "Factors Controlling the Concentration of Thirteen Rare Metals in Sea Water," in: Geochemistry of Lithogenesis. Russian translation, IL (1963).

Krauskopf, K., "Separation of Iron and Manganese during Sedimentation," in: Geochemistry of Lithogenesis. Russian translation. IL (1935).

Kritsuk, I.N., "The Role of Sorption in the Formation of Eluvial-Diluvial Dispersion Haloes," Zap. Leningr. Gorn. Inst. Vol. XLV, No. 2 (1963).

Kropachev, A.M., "Coefficients of Accumulation as indicators of the Migrational Ability of Chemical Elements in the Weathered Crust," Uch. Zap. Permsk. Gos. Univ. Vol. 14 (1959).

Krumbein, W.C., and R.M. Garrels, "Origin of Chemical Sediments and Their Classification on the Basis of pH and Redox Potentials," in: Thermodynamics of Geochemical Processes. Russian translation, IL (1960).

Kuzemkina, E.N., "Certain Secondary Processes in Mesozoic Bauxites of the Northwest Turgai," in: The Weathered Crust, Part 4. Izd. Akad. Nauk SSSR (1962).

Kuznetsov, S.I., "Biogeochemistry of Sulfur," Izv. Akad. Nauk SSSR Ser. Biol. No. 5 (1963).

Kuznetsov, S.I., M.V. Ivanov, and N.N. Lyalikova, Introduction to Geological Microbiology. McGraw-Hill, New York (1963).

Latimer, W.M., Oxidizing States of Elements and Their Potential in Aqueous Solutions. Russian translation, IL (1954).

Lebedev, V.I., Principles of Energy Analysis of Geochemical Processes. Leningr. Univ. (1957).

Leith and Mead. Metamorphic Geology. New York (1915).

Letnikov, F.A., Chemical Affinity and Its Possible Application to Geochemistry. Vestn. Akad. Nauk Kazakhstan SSR No. 12 (1963).

Letnikov, F.A., "The Isobar-Isothermal Potentials of Mineral Formation and Their Use in Geochemistry," A dissertation. Alma-Ata (1964).

Levchenko, V.M., "On the Formation of Soda in Ground Waters," in: Hydrochemical Materials, Vol. XXXI. Izd. Akad. Nauk SSSR (1961).

Lisitsin, A.K., "On the Modes of Occurrence of Uranium in Ground Waters and the Conditions of Its Precipitation as UO_2," Geokhimiya No. 9 (1962).

Lisitsin, A.K., "Characteristics of Hydrogeochemical Investigation," Geokhimiya No. 2 (1963).

Lukashev, K.I., Articles on the Geochemistry of Supergenesis. Izd. Akad. Nauk Belorussian SSR, Minsk (1963).

Lukashev, K.I., Geochemical Processes of Migration and Concentration of Elements in the Biosphere. Izd. Akad. Nauk Belorussian SSR, Minsk (1957).

Lukashev, K.I., Principles of Lithology and Geochemistry of the Weathered Crust. Izd. Akad. Nauk Belorussian SSR, Minsk (1958).

Makarenko, F.A., "Certain Common Problems in the Theory of Zoning of Ground Waters," Savarensky Laboratory for Hydrogeological Problems, Vol. 16, Izd. Akad. Nauk SSSR (1958).

Makedonov, A.V., "Certain Relationships in the Geographical Distribution of Concretions in Sediments and Soils," Izv. Akad. Nauk SSSR Ser. Geol. (1957).

Manuilova, N.S. "On Cupriferous Sandstones of Dzhezkazgan," All-Union Mineralog. Soc., 2nd Series, Vol. 83, No. 4 (1954).

Marinov, N.A., and V.N. Popov, Hydrogeology of the Mongolian Peoples Republic. Gosgeoltekhizdat (1963).

Mason, B., Principles of Geochemistry. Wiley, New York (1958).

Maximovich, G.A., Chemical Geography of Waters on the Land. Geografgiz (1955).

Mechtieva, V.L., "Distribution of Microorganisms in New and Old Clay Sediments," Tr. Inst. Mikrobiol. Akad. Nauk SSSR Vol. IX (1961).

Messineva, M.A., "Geological Activity of Bacteria and Its Effect on Geochemical Processes," Tr. Inst. Mikrobiol. Akad. Nauk SSSR Vol. IX (1961).

Methods of Investigation of Sedimentary Rocks, Vols. I and II. Gosgeoltekhizdat (1957).

Mikhailov, A.S., "Behavior of Molybdenum in the Oxidation Zone and Geochemical Exploration of Its Deposits," A dissertation. Tomsk. Politekh. Inst. (1963).

Mohr, E.C.J., and F.A. Baren, Tropical Soils. Interscience, London–New York (1954).

Narkelyun, L.F., Geology and Mineralization of the Dzhezkazgan Ore Deposits. Tr. GIGEM No. 87. Izd. Akad. Nauk SSSR (1963).

Oleinikov, A.N., "The Role of Living Matter in the Evolution of the Terrestrial Atmosphere," in: Significance of the Biosphere in Geologic Processes, Transactions of the 5th and 6th Sessions All-Union Paleontol. Soc. Gosgeoltekhizdat (1962).

"Ore-Bearing Sedimentary Formations and Ore Zoning in Artesian Petroliferous Basins of Central Asia," in: Natural Resources (1964).

Orfanidi, K.F., "Hydrochemical Zoning of Artesian Waters and Its Relationship to Climates," Dokl. Akad. Nauk SSSR Vol. 144, No. 5 (1962).

Ovchinnikov, A.M., General Hydrogeology. Gosgeoltekhizdat (1955).

Ovchinnikov, A.M., "Hydrogeological Conditions of Hydrothermal Processes," Byul. Mosk. Obshchestva Ispytatelei Prirody Otd. Geol. Vol. XXXII (1957).

Ovchinnikov, A.M., "The Value of Geochemistry and Hydrogeology in the Search for Ore Deposits," Interdepartmental Conference on Hydrochemical Methods of Exploration, Tomsk Univ. (1962).

Ovchinnikov, A.M., "The Partition of Territories on Hydrogeological Regions as the Basis of the Rational Development of Geochemical Methods of Exploration for Ores," Conference on Hydrogeology and Engineering Geology, Geochemical Section (1963).

Ovchinnikov, A.M., Mineral Waters. Gosgeoltekhizdat (1963).

Pallman, H., "Die Bodentypen der Schweiz," Mitt. Gebiete Lebensm. Hyg. Vol. XXIV (1933).

Parfenova, E.I., and E.A. Yarilova, Mineralogical Investigations in Pedology. Izd. Akad. Nauk SSSR (1962).

Peine, Ya. V., Biochemistry of Soils. Sel'khozgiz (1961).

Perel'man, A.I., "An Experiment in Energy Characterization of Reaction of Chemical Weathering," A dissertation. Mosk. Gos. Univ. (1941).

Perel'man, A.I., "An Experiment in Application of Thermochemical Methods in Pedology," Pochvovedenie No. 5 (1947).

Perel'man, A.I., Catagenesis. Izv. Akad. Nauk SSSR. Ser. Geol. No. 8 (1959a).

Perel'man, A.I., "Certain Problems in the Geochemistry of Sedimentary Ore Deposits of the Type of Cupriferous Sandstones," Tr. GIGEM No. 28. Izd. Akad. Nauk SSSR (1959b).

Perel'man, A.I., "Processes of Saline Migration on the Neogene Plains of Eastern Turkmenia and Western Uzbekistan," Tr. GIGEM No. 25. Izd. Akad. Nauk SSSR (1959c).

Perel'man, A.I., Geochemistry of Epigenesis. Izd. Vysshaya Shkola (1961a).

Perel'man, A.I., The Geochemistry of Landscapes. Geografgiz (1961b).

Perel'man, A.I., "Occurrence of Geochemical Types of Weathered Crust in Terrigenous Sediments of the USSR," in: The Weathered Crust, Part 5. Izd. Akad. Nauk SSSR (1963a).

Perel'man, A.I., "Certain Peculiarities of Aqueous Migration of Chemical Elements in Landscapes," in: Problems of Geochemistry, Part 5. Tr. GIGEM No. 99. Izd. Akad. Nauk SSSR (1963b).

Perel'man, A.I., and S.G. Batulin, "Migrational Series of Elements in the Weathered Crust," in: The Weathered Crust, Part 4. Izd. Akad. Nauk SSSR (1962).

Perel'man, A.I., and E.N. Borisenko, "Notes on the Geochemistry of Copper in the Supergene Zone," in: Problems of Geochemistry. Tr. GIGEM No. 70. Izd. Akad. Nauk SSSR (1962).

Petrov, N.P., "Conditions of Formation of a Sulfur Deposit in the Southwestern Foothills of Ghissar," Report of the Uzbek Department of the All-Union Mineralog. Soc. Vol. VIII (1955).

Pokrovskii, V.A., "The Lower Limit of the Biosphere in the European USSR, as Defined by New Geothermal Investigations," Tr. Inst. Mikrobiol. Akad. Nauk SSSR Vol. IX (1961).

Polikarpochkin, V.V., Geochemical Exploration for Ore Deposits by Means of Dispersion Flows, Sov. Geol. No. 4 (1962).

Polyakov, V.A., A.I. Panteleev, and I.N. Vorob'eva, "Investigation of the Role of Sorption in Migration of Microelements in Ground Waters in Connection with the Development of Hydrochemical Methods of Exploration for Ores," in: Materials, Technical Reports of the XVII Hydrochemical Conference. Novocherkassk (1963).

Polynov, B.B., The Weathered Crust (1934), in: Selected Works. Izd. Akad. Nauk SSSR (1956).

Polynov, B.B., The Red Weathered Crust and Its Soils (1944), in: Selected Works, Izd. Akad. Nauk SSSR (1956).

Polynov, B.B., Contemporary Problems in the Theory of Weathering. Izv. Akad. Nauk SSSR Ser. Geol. No. 2 (1944).

Polynov, B.B., "Geochemical Landscapes," in: Geographical Studies. Geografgiz (1952).

Polynov, B.B., "On the Geological Role of Organisms," in: Problems of Geography No. 33. Geografgiz (1953).

Ponov, A.B., Editor, Geochemistry of Lithogenesis. Russian translation, IL, Moscow (1963).

Popov, I.V., "Principles of Geological Engineering in Soils," in: Transactions of the State Trust for Specialized Geological Mapping, Part 9. Gosgeoltekhizdat (1941).

Popov, V.I., and A.L. Vorob'ev, "Certain Mineral and Geochemical Features of Terrigenous Formations in Deserts," Report of the Uzbek Department of the All-Union Mineralog. Soc. Vol. VIII (1955).

Popov, V.I., "On Differentiation among Syngenesis, Epigenesis and Metamorphism of Sedimentary Rocks," Izv. Akad. Nauk SSSR Ser. Geol. No. 1 (1957).

Popov, V.I., S.D. Makarova, Yu. V. Stankevich, and A.D. Filippov, Manual for the Determination of Sedimentary Facies and Methods of Paleogeological Mapping. Gosgeoltekhizdat (1963).

Popov, V.M., "Diagenetic and Epigenetic Phenomena in Cupriferous Sandstones of the Donets Basin," Izv. Akad. Nauk Kirg. SSR No. XI (1954).

Popov, V.M., "Certain Specific Features of the Geology of the Cupriferous Sandstones of Central Kazakhstan," Tr. Geol. Inst. Akad. Nauk Kirg. SSR No. 6 (1955).

Prescott, I.A., and R.L. Pendleton, Laterite and Lateritic Soils. London (1952).

Priklonsky, V.A., and N.A. Oknina, "Diffusional Processes in Argillaceous Rocks and Their Significance in Hydrogeology and Engineering Geology," in: Problems of Hydrogeology. Gosgeoltekhizdat (1960).

"Problems of Geochemistry of Ground Water," in: Natural Resources (1964).

Pustovalov, L.V., Petrography of Sedimentary Rocks. Gosgeoltekhizdat (1940).

Pustovalov, L.V., "Secondary Alteration of Sedimentary Rocks and Its Geological Significance," in: Secondary Alteration of Sedimentary Rocks. Izd. Akad. Nauk SSSR (1956).

Rakovskii, A.A., Treatise on Physical Chemistry. Mosk. Gos. Univ. (1939).

Rankama, K., and T.G. Sahama, Geochemistry, 3rd ed. U. of Chicago (1955).

Rasskazov, N.M., "Hydrogeochemistry and Certain Peculiarities of Ground Waters of the Kirs Range (Western Sayany)," A dissertation. Tomsk. Gos. Univ. (1963).

Rateev, M.A., "The Role of the Climate and Tectonics in the Genesis of Sedimentary Clay Minerals," Transact. Clay Inst. OGGN. Izd. Akad. Nauk SSSR (1960).

Razumova, V.N., "Cretaceous and Tertiary Formations of the West-Central and Western Part of Southern Kazakhstan," Tr. Geol. Inst. Akad. Nauk SSSR No. 46 (1961).

Rode, A.A., The Podzol Forming Process. Izd. Akad. Nauk SSSR (1937).

Rodionov, N.V., Karst of the European Part of the USSR, the Urals, and the Caucasus. Gosgeoltekhizdat (1963).

Rozhkova, E.V., O.V. Scherbak, and V.M. Saakyan, "The Role of Sorption in the Concentration of Zinc in Sedimentary Rocks," in: Mineral Materials, Part 6. Gosgeoltekhizdat (1962).

Rukhin, L.B., Principles of Lithology, 2nd ed. Gosgeoltekhizdat (1961).

Safronov, N.I., Principle of Geochemical Methods of Exploration for Mineral Deposits. VITR, Leningrad (1962-1963).

Salai, A., "The Significance of Humus in the Geochemical Enrichment in Uranium," Transact. Int. Conf. on Peaceful Uses of Atomic Energy (Geneva, 1958), Vol. 8: The Geology of Atomic Materials. Atomizdat (1959). English edition published by Pergamon, London—New York.

Samodurov, P.S., "Lithologic and Geochemical Characteristics and Paleogeographic Conditions of Formation of Loess and Loessoids in Belorussia," A dissertation. Minsk (1963).

Sapozhnikov, D.G., A Contribution to the Theory of Investigation of Sedimentary Ore Deposits. Izd. Akad. Nauk SSSR (1961).

Saukov, A.A., "Causes Delimiting the Number of Minerals," in: Problems of Mineralogy, Petrography, and Geochemistry. Izd. Akad. Nauk SSSR (1946).

Saukov, A.A., Geochemistry. Gosgeoltekhizdat (1951).

Saukov, A.A., "On the Development of the Migrational Factors of Chemical Elements," Priroda No. 2 (1958).

Saukov, A.A., "Some Comments on Hydrothermal Solutions and Hydrothermal Ore Deposits," in: Problems of Geochemistry. Tr. GIGEM No. 46. Izd. Akad. Nauk SSSR (1961).

Saukov, A.A., Geochemical Methods of Investigating of Mineral Deposits. Mosk. Gos. Univ. (1963).

Shcherbakov, A.V., "Geochemical Criteria of Redox Environments in the Subsurface Hydrosphere," Sov. Geol. No. 56 (1956).

Shcherbakov, A.V., "The Role of Exogene Metamorphism in the Formation of Natural Gases in the Lithosphere and the Subsurface Hydrosphere," in: Problems of Hydrogeology and Engineering Geology, No. 16. Gosgeoltekhizdat (1959).

Shcherbakov, A.V., "The Principal Paleo-Hydrogeochemical Features of the Greater Donbass," in: Geochemistry of Ground Waters of Certain Regions of the European USSR. Izd. Akad. Nauk SSSR (1963).

Shcherbina, V.V., Geochemistry. Izd. Akad. Nauk SSSR (1939).

Shcherbina, V.V., Glauberite, Glauberitic Rocks, and Their Weathered Crust. Izd. Akad. Nauk Kirg. SSR, Frunze (1952).

Shcherbina, V.V., "A Contribution to the Geochemistry of Oxidation Zones of Ore Deposits," Sov. Geol. No. 43 (1955).

Shcherbina, V.V., Chemistry of the Mineral Forming Processes in Sedimentary Rocks, Vols. 3 and 4. Mosk. Gos. Univ. (1956).

Shcherbina, V.V., "Behavior of Uranium and Thorium in Conditions of the Sulfate–Carbonatic and the Phosphatic Media of the Supergene Zone," Geokhimiya No. 6 (1957).

Shcherbina, V.V., Editor, Thermodynamics of Biochemical Processes. Russian translation from English and German, IL (1960).

Shcherbina, V.V., "Behavior of Certain Rare and Disperse Elements in the Supergene Zone," Sov. Geol. No. 6 (1962a).

Shcherbina, V.V., "Methods of Investigation of Migration of Chemical Elements During Chemical Processes," Geokhimiya No. 11 (1962b).

Shcherbina, V.V., and L.I. Ignatova, "New Data on the Geochemistry of Copper in the Supergene Zone," Trans. All-Union Mineralog. Soc. 2nd series, Vol. 84, No. 3 (1955).

Shvartsev, S.L., "Some Results of Geochemical Investigation in the Permafrost Regions," Geol. Rudn. Mestorozhd. No. 2 (1963).

Shvedas, A., "A Study of Factors Controlling the Behavior of Copper in Soils of the Lithuanian SSR," in: Microelements in Agriculture and Medicine, Reports of the Fourth All-Union Conference, Ukr. Acad. Ag. Sci. Izd. UASKhN, Kiev (1963).

Sidorenko, A.V., "Principal Features of Mineral Formation in Deserts," in: Problems of Sedimentary Mineralogy, Vols. 3 and 4. L'vov. Gos. Inst. (1956).

Sidorenko, A.V., "Caliches of Mexican Deserts," Izv. Akad. Nauk SSSR Ser. Geol. No. 1 (1958).

Sidorenko, A.V., "Calcareous Crusts in the Egyptian Desert," Dokl. Akad. Nauk SSSR Vol. 128, No. 4 (1959).

Silin-Bekchurin, A.I., "Conditions of Formation of Salt Waters in Arid Zones," in: Problems of Saline Contamination of Soils and Springs. Izd. Akad. Nauk SSSR (1960).

Smirnov, A.A., "The Role of Ground Water in the Formation, Preservation, and Destruction of Organic Gas Deposits," in: Problems of Hydrogeology and Engineering Geology No. 14. Gosgeoltekhizdat (1956).

Smirnov, S.I., "Geochemistry of Ground Water in the Supergene Zone of Sulfide Ore Deposits," in: Problems of Hydrogeology. Gosgeoltekhizdat (1960).

Smirnov, S.S., The Oxidation Zone of Sulfidic Ore Deposits. Izd. Akad. Nauk SSSR (1955).

Smyshlyaev, S.I., and N.P. Edeleva, "Determination of Mineral Solubilities," in: Chemistry and Chemical Technology, No. 6. Izd. Vysshikh Uchebn. Zavedenii (1962).

Sokolov, A.S., "Geological Factors of Structure and Distribution of Sulfur Deposits," Trans. Inst. of Raw Chemical Materials, Part 5. Goskhimizdat (1959).

Sokolov, D.S., Fundamentals of the Karst Development. Gosgeoltekhizdat (1962).

Sokolov, I.A., and T.A. Sokolova, "On the Zoned Type of Soils in Permafrost Regions," Pochvovedenie No. 10 (1962).

Solovov, A.P., Fundamentals of the Theory and Practice of Metallometric Surveys. Alma-Ata, Izd. Akad. Nauk Kaz. SSR (1959).

Strakhov, N.M., "Stages in the Development of the Outer Geospheres and the Formation of Sedimentary Rocks during the Earth's History," Izv. Akad. Nauk SSSR Ser. Geol. No. 12 (1962).

Strakhov, N.M., Fundamentals of the Theory of Lithogenesis, Vols. I-III. Izd. Akad. Nauk SSSR (1960-1963). English translation, Consultants Bureau, New York, Vol. 1 (1967); Vols. II and III in preparation.

Strakhov, N.M., Lithogenetic Types and Their Evolution through the Earth's History. Gosgeoltekhizdat (1963).

Sukharev, G.M., and M.V. Miroshnikov, Ground Waters of the Caucasian Oil and Gas Deposits. Gostoptekhizdat (1963).

Sulin, V.A., Formation, Classification, and Composition of Ground Waters, Part I. Izd. Akad. Nauk SSSR, Moscow—Leningrad (1948).

Sveshnikov, G.B., "Electrochemical Leaching of Sulfidic Ores and Its Role in the Formation of Aqueous Dispersion Haloes of Heavy Metals," in: Geological Results of Applied Geochemistry and Geophysics. Gosgeoltekhizdat (1960).

Swineford, A., Editor, Clays and Clay Minerals: Proceedings of the National Conference on Clays and Clay Minerals. Pergamon, London—New York (1960).

Tatarinov, P.M., Conditions of Formation of Mineral Deposits. Gosgeoltechizdat (1963).

Tatarskii, V.B., "De-dolomitization and Its Problems," Vestn. Leningr. Univ. No. 1 (1953).

Tkachuk, V.G., "Mineral Waters of the Southern Part of Eastern Siberia, Their Origin and Utilization," in: Problems of Hydrogeology. Gosgeoltekhizdat (1960).

Tugarinov, A.I., "Factors of Formation of Metallogenic Provinces," in: Chemistry of the Earth's Crust, Vol. 1. Izd. Akad. Nauk SSSR (1963).

Udodov, P.A., et al., An Experiment in Geochemical Prospecting in Siberia. Vysshaya Shkola (1962).

Uklonsky, A.S., The Paragenesis of Sulfur and Petroleum. Izd. Akad. Nauk Uzbek SSR. Tashkent (1940).

Valyashko, M.G., "Geochemistry of Halogenesis," Collection of Works, Geol. Dept. Moscow Univ. Izd. Mosk. Gos. Univ. (1961).

Valyashko, M.G., Geochemical Relationships in the Formation of Potash Deposits. Izd. Mosk. Gos. Univ. (1962).

Van Hise, C.B., Treatise on Metamorphism. USGS, Washington (1904).

Verigina, K.V., "On Processes of Gley Formation in Soils," Tr. Pochv. Inst. Akad. Nauk SSSR Vol. X (1953).

Verigina, K.V., "On the Contents of Zinc, Copper, and Cobalt in Some Soils," in: Microelements in Agriculture and Medicine, Fourth All-Union Conference, Ukr. Acad. Ag. Sci. Izd. UASKhN, Kiev (1962).

Vernadskii, V.I., History of Minerals of the Earth's Crust, Vol. 1, Part 1. Izd. Akad. Nauk SSSR (1925).

Vernadskii, V.I., Selected Works, Vols. I-V. Izd. Akad. Nauk SSSR (1954-1960).

Vernadskii, V.I., and S.M. Kurbatov, Terrestrial Silicates and Aluminosilicates and Their Analogues. ONTI (1937).

Vinogradov, A.P., Geochemistry of Living Matter. Izd. Akad. Nauk SSSR (1932).

Vinogradov, A.P., Geochemistry of Rare and Dispersed Chemical Elements in Soils. Consultants Bureau, New York (1959).

Vinogradov, A.P., "The Average Content of Chemical Elements in the Principal Types of Igneous Rocks of the Earth's Crust," Geokhimiya No. 7 (1962).

Vinogradov, A.P., "Biogeochemical Provinces and Their Role in Organic Evolution," Geokhimiya No. 3 (1963).

Volkova, O. Yu., and A.I. Germanov, "New Data on the Biogeochemistry of the Supergene Migration of Chemical Elements," in: Problems of Geochemistry, Part V. IGEM. No. 88. Acad. Sci. USSR (1963).

Weathered Crust, The, Parts I-V. Izd. Akad. Nauk SSSR (1952-1963).

Wolfson, F.I., Problems in the Investigation of Hydrothermal Ore Deposits. Gosgeoltechizdat (1962).

Yarilova, E.A., "Investigations of Manganese Migration in Soils," Tr. Pochv. Inst. Akad. Nauk SSSR Vol. XXIV (1940).

Yurkevich, I.A., Studies of the Methods of the Facies and Geochemical Investigation of Sedimentary Rocks. Izd. Akad. Nauk SSSR (1958).

Zaitsev, I.K., and N.I. Tolstikhin, "Basic Features of Hydrogeology of the USSR," in: Problems of Hydrogeology. Gosgeoltekhizdat (1960).

Zelenova, O.I., "Epigenetic Distribution of Iron in Red Rocks of Kul'dzhuktau," in: Problems of Geochemistry, Tr. GIGEM No. 99. Izd. Akad. Nauk SSSR (1963).

Zhuravleva, E.G., "Content of Zinc, Copper and Cobalt in Some Soils of the Chita State," in: Microelements in Agriculture and Medicine, Fourth All-Union Conference, Ukr. Acad. Ag. Sci. Izd. UASKhN, Kiev (1962).

Zul'fugarlii, D.I., Distribution of Microelements in Caustobiolites, Organisms, Sedimentary Rocks, and Ground Waters. Izd. Azerb. Gos. Univ., Babu (1960).

Index